FOSTERING COMPUTATIONAL THINKING AMONG UNDERREPRESENTED STUDENTS IN STEM

This book broadly educates preservice teachers and scholars about current research on computational thinking (CT). More specifically, attention is given to computational algorithmic thinking (CAT), particularly among underrepresented K–12 student groups in STEM education.

Computational algorithmic thinking (CAT)—a precursor to CT—is explored in this text as the ability to design, implement, and evaluate the application of algorithms to solve a variety of problems. Drawing on observations from research studies that focused on innovative STEM programs, including underrepresented students in rural, suburban, and urban contexts, the authors reflect on project-based learning experiences, pedagogy, and evaluation that are conducive to developing advanced computational thinking, specifically among diverse student populations.

This practical text includes vignettes and visual examples to illustrate how coding, computer modeling, robotics, and drones may be used to promote CT and CAT among students in diverse classrooms.

Jacqueline Leonard is Professor of Mathematics Education in the School of Education at the University of Wyoming, USA.

Jakita O. Thomas is the Philpott Westpoint Stevens Associate Professor of Computer Science and Software Engineering at Auburn University, USA.

Roni Ellington is Associate Professor of Mathematics Education in the Department of Advanced Studies Leadership and Policy at Morgan State University, USA.

Monica B. Mitchell is founder and President of MERAssociates, LLC (MERA), an award-winning evaluation consultancy based in the greater metropolitan area of Washington, D.C., USA.

Olatokunbo S. Fashola is Research Professor and the Faculty Coordinator for the Dual Enrollment Program in the School of Education at American University in Washington, D.C., USA.

FOSTERING COMPUTATIONAL THINKING AMONG UNDERREPRESENTED STUDENTS IN STEM

Strategies for Supporting Racially Equitable Computing

Jacqueline Leonard, Jakita O. Thomas, Roni Ellington, Monica B. Mitchell, and Olatokunbo S. Fashola

Routledge
Taylor & Francis Group

NEW YORK AND LONDON

First published 2022
by Routledge
605 Third Avenue, New York, NY 10158

and by Routledge
2 Park Square, Milton Park, Abingdon, Oxon, OX14 4RN

Routledge is an imprint of the Taylor & Francis Group, an informa business

© 2022 Taylor & Francis

The right of Jacqueline Leonard, Jakita O. Thomas, Roni Ellington, Monica B. Mitchell, and Olatokunbo S. Fashola to be identified as authors of this work has been asserted by them in accordance with sections 77 and 78 of the Copyright, Designs and Patents Act 1988.

Library of Congress Cataloging-in-Publication Data
Names: Leonard, Jacqueline, author. | Thomas, Jakita, 1977- author. |
Ellington, Roni M., author. | Mitchell, Monica B., author. | Fashola,
Olatokunbo S., author.
Title: Fostering computational thinking among underrepresented students in
STEM : strategies for supporting racially equitable computing /
Jacqueline Leonard, Jakita O. Thomas, Roni Ellington, Monica B.
Mitchell, and Olatokunbo S. Fashola.
Description: New York, NY : Routledge, 2021. | Includes bibliographical
references and index.
Identifiers: LCCN 2021002108 | ISBN 9780367456504 (hbk) | ISBN
9780367456511 (pbk) | ISBN 9781003024552 (ebk)
Subjects: LCSH: Computer science--Study and teaching (Elementary)--United
States. | Minorities--Education--United States. | Educational
equalization--United States. | Computer literacy--United States. |
Digital divide--United States.
Classification: LCC QA76.27 .L46 2021 | DDC 004.071--dc23
LC record available at https://lccn.loc.gov/2021002108

ISBN: 978-0-367-45650-4 (hbk)
ISBN: 978-0-367-45651-1 (pbk)
ISBN: 978-1-003-02455-2 (ebk)

Typeset in Bembo
by SPi Technologies India Pvt Ltd (Straive)

IN MEMORY OF
Dr. Karen D. King
Dancing with Mathematics

With advanced mathematics courses under your belt,
Your mathematical journey began at Eleanor Roosevelt.
Prowess was recognized at Spelman College
Where you graduated early to advance new knowledge.

Voice undeniable you built *Terp* reputation,
Niche carved in pure math will influence generations.
PhD in '97—you became a rising star.
San Diego, East Lansing, New York—you continued to raise the bar.

From the professoriate to the National Science Foundation,
Mentor, advisor, colleague, friend—worthy of commendation.
How should we honor 48 years and define your legacy?
Dancing with mathematics goes down in history.

Jacqueline Leonard
January 2020

CONTENTS

FIGURES

TABLES

FOREWORD

TOWARD RACIALLY EQUITABLE COMPUTING EDUCATION

It is a great honor for me to write this foreword. As an immigrant scholar of color in an intensely White academy, I find myself in deep solidarity with the authors. As my colleagues and I have argued recently (Philip & Sengupta, 2020), the fields of computing, STEM education, and the Learning Sciences need to imminently reorganize themselves in order to dismantle monuments of epistemic violence that in turn are erected by the incessant tyranny of the *White gaze*. I see this book as a *gift* in terms of offering us a path forward for countering this tyranny in these fields.

But what do we mean by the *White gaze*? We can begin with Toni Morrison's poignant rebuttal of the critiques of her novels for not centering White protagonists; in a 1998 interview (as quoted in Paris, 2019), she stated: they act "as though our (Black) lives have no meaning and no depth without the White gaze."[1] In the White gaze, the richness of Black lives in Morrison's work was deemed insufficient or even deficient for the relative lack of focus on White protagonists. In our context, under the omnipresent White gaze, bodies of Black, Indigenous, and People of Color (BIPOC) are positioned as "a spectacle of performances" by White and privileged researchers, who in turn "are fixated on 'saving' the disenfranchised…" (Harrington et al., 2019, as cited in Rankin et al., 2020, p. 69).

I must also admit here that my reading of this book is shaped by my own position as a critical phenomenologist of computing. Given this background, this book offers several important insights, which I have tried to synthesize here. First, I position the book in light of the troubled history of computing and technoscience, which has historically decentered BIPOC. Then, I reflect directly on the chapters, and highlight three key emergent themes that are epistemological and methodological commitments for thwarting the White gaze and centering the voices and experiences of Black, Latinx, and Indigenous children and youth. I conclude with a return to solidarity to emphasize the value of recognizing cultural and computational heterogeneity in re-orienting computing education toward racially equitable futures.

The Troubled Roots of Computing

The relationship of computing with color is treacherous at best, and violent and oppressive at its worst. Computing education, unsurprisingly, has left out BIPOC, as evident in the inadequate and unequal opportunities for computing in predominantly Black and Latinx classrooms in the United States. This relationship is further exacerbated by the perceived ethical neutrality of technology, despite growing concerns that many *technocorrections* (Hatch & Bradley, 2017) affect BIPOC disproportionately in oppressive ways. The danger of such ahistorical, uncritical, and neutral positionings of technology (and discipline in general) lies in perpetuating and naturalizing epistemic violence in the form of "Indigenous erasure, anti-blackness, and of enabling and glorifying extraction paradigms to produce unsustainable technological futures" (Bang, 2020, p. 435).

Abstractions have taken on a central and defining role in what has now come to be known as computational thinking. As we have argued elsewhere (Sengupta et al., 2021), learning to code has now become synonymous with mastering abstractions. Historically, in the early 1900s, Lewis Terman, a pioneer of IQ testing, declared that Black people and immigrants of color in the U.S. are not capable of mastering abstractions and should instead be prepared for becoming efficient workers (Terman, 1925). Although this proclamation is no longer viewed favorably in education research, the disparity in computing infrastructure and opportunities for Black, Latinx, and Indigenous students in the U.S. continues to persist. It is also noteworthy that the notion of abstractions is a greatly reduced representation of the complexity of our experiences of code and computing. Thus, given the distinctly anti-Black and racist history of computing and technoscience, it becomes all too important to acknowledge and celebrate the fact that Black, Latinx, and Indigenous children are indeed capable and brilliant cognitive actors in computing. It is in these troubled waters of epistemic violence and erasure of BIPOC that the authors launch this book, directing us toward strategies for making computing education racially equitable.

Advancing Racial Equity in Computing

With a growing recognition of systemic racism and intersectional inequities in praxis and research in STEM and computing education, we are now faced with the necessary task of countering these forces in our research and practice. There are, of course, fundamental critiques of discipline that must also be recognized in computing and STEM education, which include acquiescence to imperialist and nationalist ideologies within the curricula and classroom discourse and an instru-mentalist approach that positions STEM and computing education as means for "workforce readiness." But the authors of this book remind us that while we need to work toward addressing these more fundamental epistemological and ontological issues—a long term outlook—we also need to work toward addressing the existing racial inequalities in the technological workforce—a more immediate concern. The authors remind us that the White gaze is insuf-ficient, and we need to think carefully about the "how," the "for what," the "for whom," and the "with whom," as we commit to culturally responsive and critical pedagogies (Philip et al., 2018).

In Chapters 1 and 2, the term intersectionality is introduced to describe a theoretical position that promotes a deeper understanding of the multi-dimensionality inherent in racial exclusion. This means being able to recognize and value the simultaneous, multiple positionalities of Black students—who, for example, may also be women and disabled—as well as recognizing that Black role models may also have multiple positionalities. For example, Bessie Coleman, who features prominently in this book at the center of several learning activi-ties as a role model, was the first woman of African-American descent, and also the first woman of Native-American descent, to hold a pilot's license in the U.S. Moreover, Chapter 2 centers the computational game design experiences of Black girls in sixth grade, while Chapter 5 centers voices of Indigenous stu-dents in the Arapaho community around the role model of Bessie Coleman. And Chapter 7 is a necessary reminder that paying attention to intersectionality must also inform the design of evaluation in computing education, which needs to recognize the culturally responsive nature of teaching, as well as the het-erogeneous nature of culturally responsive, computational experiences of Black, Latinx, and Indigenous students. Central to this, however, is a necessary move away from technocentrism.

Beyond Technocentrism and Device-Centered Approaches

Technocentrism (Papert, 1987) refers to the fallacy of referring all questions about technology to the technology itself, rather than trying to understand the experiences of and infrastructures around it. Emphasis on heterogeneity is critical for avoiding technocentric tropes in computing and STEM education (Papert, 1987). In a truly fundamental sense, this book illustrates that paying attention to heterogeneity is key for designing and supporting culturally responsive comput-ing and countering the White gaze that is also centered in technocentric com-puting education.

The authors offer us counternarratives of the success of Black, disabled, and marginalized students of color and their teachers while also pointing out the frailties of White ideologies that underlie many of the paradigmatic approaches in computing education. As Ms. Davis, a teacher, says in Chapter 6 of this text:

"There is no deficit in our children ... and they honestly always succeed" (p. 110). And yet, a striking (and understated) finding in Chapter 3 is that the underrepresented students persisted in activities such as the Scalable Game Design, engaging deeply with the activities and developing expertise in coding, mathematical reasoning, and game design despite the lack of cultural responsiveness in the activities. The authors also remind us that Black, Latinx, and Indigenous children in the U.S., who have been ignored and oppressed over generations, must be provided with technological infrastructure that allow them to illustrate unique aspects of their cultures and communities.

The research on underrepresented students' perceptions of and experiences in science and technology in informal learning environments is sparse. Chapters 4 and 5 both seek to fill this gap. Chapter 4 offers us images and insights from afterschool engagements with predominantly Black students and teachers in Denver and Philadelphia. The authors illustrate that Black youth, who have been underserved in STEM, came to see computer modeling and 3D printing as ways and tools to take on real-world problems facing their communities. While the students' minds may have been at work with the computer, their hearts were centered in their communities—an inextricable duality that is resonant throughout the chapters, and a reality that we, as researchers, must learn to center in our work.

Another important finding in Chapter 4 is the importance of role models for marginalized students. Researchers and educators must always remember that the historical and systemic nature of epistemic violence is inscribed deep within public education: that is, the debilitating part of feeling marginalized is not merely limited to the experience of being left out or erased in the moment but the persistence of this erasure over generations. What is needed then, as the authors remind us here, are opportunities of solidarity with role models. Being able to listen to Capt. Ed Dwight's experiences with racism as the first Black astronaut trainee is one such example, which offered students not only narratives of hope—what is possible—but also words and stories of resilience. Similarly, in Chapter 6, the authors inform us that professional development for teachers must also integrate culturally responsive, equity-centered perspectives that focus on assets-based narratives of children of color that in turn are grounded in and centered in the teachers' and students' real-world experiences.

Taken together, the chapters in this book offer a fundamental challenge to device-centered imaginations of computing education, which emphasize technical productions over the complexity of human experience. Computing education in this White gaze positions children of color as deficient rather than resilient, fundamentally devaluing their lived experiences and friendships outside the computer. To this end, an important contribution of this book is the reminder that thwarting this gaze requires explicit efforts to decolonize computing education.

Decolonizing Computing Education

In Chapter 5, the authors take us deep into decolonizing computing education by foregrounding Indigenous epistemologies and pedagogical approaches. The authors report on a collaboration with Native-American teachers who lived and worked on the Wind River Indian Reservation (WRIR) in Wyoming. Their work was grounded in Tribal Critical Theory (Brayboy, 2005), Indigenous perspectives that center place-based education, and culturally specific pedagogy. While designing learning environments must be intentional to a certain degree on the part of the designers, teachers, and researchers, decolonizing computing education necessitates a specific form of intentionality: to center the voices and values of Indigenous communities by enlivening connections to the place—land, air, and water—as well as ways of knowing and being that are learned and taught intergenerationally through stories as well as through centering reverence and respect for sacred lands.

One of the most profound findings in this chapter is that families of multi-aged siblings worked together—and talked together—as they engaged collectively in computational participation. They recorded drone images of water, land, and plants. Young women created collective STEM identity artifacts by documenting their kinship in the form of group photographs, and young men celebrated their skateboarding identity with drone tracking and video recording. While emphasis on the individual child is another hallmark of White, upper-class ideology that has shaped much of computing education, the authors offer a strikingly different, decolonized account in which learners—and not technologies—are centered through collective and community-level engagement.

The authors, thus, offer rich images of *computational heterogeneity* (Sengupta et al., 2021) as evident in intergenerational and collective engagement in complex projects. Code leaves the flattened digital realm and instead is enlivened through drawing maps on animal hide, interweaving of computational representations and Arapaho language, recording images and videos in reverence to the ancestors and the land, developing identity artifacts together, and so on. The connection between the heterogeneity of language is central to understanding the complexity of the experience of computing—and a commitment to myriad ways of knowing in Indigenous perspectives.

Epilogue: Re-Orienting Computing Away from the White Gaze

The authors of this book call for racially equitable computing education that is grounded in intersectionality, culture, and place. This in turn requires looking beyond the White gaze and device-centered approaches in computing education—beyond computational artifacts and the technocentric measures of computational thinking—and instead, centering the heterogeneity of multimodal, cultural, and historical forms of knowing and being that are central to

BIPOC and their communities. In doing so, the authors argue that we can center culturally responsive, community-centered, and land-based approaches in our research so that technology becomes a place for deepening our intergenerational relationship with each other and the world around us.

Pratim Sengupta, PhD
Professor, Learning Sciences
Werklund School of Education, University of Calgary

Note

1 Morrison, T. (1998). From an interview on Charlie Rose. Public Broadcasting Service. http://www.youtube.com/watch?v¼F4vIGvKpT1c

References

Bang, M. (2020). Learning on the move toward just, sustainable, and culturally thriving futures. *Cognition and Instruction, 38*(3), 434–444.

Brayboy, B. M. J. (2005). Toward a tribal critical race theory in education. *The Urban Review, 37*(5), 425–446.

Hatch, A. R., & Bradley, K. (2017). Prisons matter: Psychotropics and the trope of silent technocorrections. In V. Pitts-Taylor (Ed.), *Mattering: Feminism, science, and materialism* (pp. 224–244). New York University Press.

Papert, S. (1987). Information technology and education: Computer criticism vs. techno-centric thinking. *Educational Researcher, 16*(1), 22–30.

Paris, D. (2019). Naming beyond the white settler colonial gaze in educational research. *International Journal of Qualitative Studies in Education, 32*(3), 217–224.

Philip, T. M., Bang, M., & Jackson, K. (2018). Articulating the "how," the "for what," the "for whom," and the "with whom" in concert: A call to broaden the benchmarks of our scholarship. *Cognition and Instruction, 36*(2), 83–88. https://doi.org/10.1080/0737 0008.2018.1413530

Philip, T. M., & Sengupta, P. (2020). Theories of learning as theories of society: A contrapuntal approach to expanding disciplinary authenticity in computing. *Journal of the Learning Sciences.* https://doi.org/10.1080/10508406.2020.1828089

Rankin, Y. A., Thomas, J. O., & Joseph, N. M. (2020). Intersectionality in HCI: Lost in translation. *Interactions, 27*(5), 68–71.

Sengupta, P., Dickes, A., & Farris, A. V. (2021). *Voicing code in STEM: A dialogical imagination.* The MIT Press.

Terman, L. M. (1925). *Genetic studies of genius, Vol 1.: Mental and physical traits of a thousand gifted children.* Stanford University Press.

ACKNOWLEDGMENTS

We are indebted to several individuals who were instrumental in guiding our work on this project. We express our sincerest thanks to Pratim Sengupta, University of Calgary, for writing the foreword to this book. Moreover, we recognize the tireless efforts of the Bessie Coleman Project team: Andrea C. Burrows, Alan Buss, Roni Ellington, Ruben Gamboa, Brandon S. Gellis, W. J. Jordan, Diane Jass Ketelhut, Mikhail Miller, Geeta Verma, and Christopher G. Wright. Additionally, we appreciate the contributions of Cara Djonko-Moore and Dan Blustein, assistant professors at Rhodes College, and Angela Jamie, Vice-Dean Indigenous, College of Arts & Science, University of Saskatchewan. We also acknowledge graduate students in the EMAT doctoral program at the University of Wyoming for reading and offering critique on the Indigenous chapter: Luke Audette, Libni Berenice Castellón, Ariane Eicke, Michael Gundlach, Kelly Hawkinson, Amy Kassel, Geoff Krall, Aylin Márquez, Karla Valesca Matute Colindres, Megan Rourke, and Angela Schanke. Furthermore, we appreciate the diligence of Marzetta Alexander for her stellar editorial service, and Peter Hill for developing the Tinkercad modules referenced in this work. We also thank the following research assistants at MER Associates for contributing to the summative evaluation: Tinielle Carson, Eileen Nolan, Keturah Postell, and Samiyah Zaman. We are also indebted to our STEM speakers, Mark Todd Clementz, Willie L. Daniels, Ed Dwight, Ron Oliversen, and Wallace Ulrich. Finally, we acknowledge the students, teachers, and administrators in Albany County School District #1 (Wyoming); Fremont County School District #25 and #38 (Wyoming); William Penn School District (Pennsylvania); Evanston Youth Club for Boys & Girls, and the Boys & Girls Clubs in Cheyenne, Wyoming, and Denver, Colorado.

The material presented in this book is supported by the Fulbright Program and the National Science Foundation (grant numbers 1311810 and 1757976).

Any opinions, findings, and conclusions or recommendations expressed in this publication are those of the authors and do not necessarily reflect the views of the Fulbright Program or the National Science Foundation.

Professor Jacqueline Leonard
School of Teacher Education, University of Wyoming
Principal Investigator, Bessie Coleman Project,
National Science Foundation (2018–2021)

BIOGRAPHICAL NOTES

Jacqueline Leonard, PhD, is Professor of STEM Education at the University of Wyoming (UW) in Laramie, WY, United States. Dr. Leonard was also promoted to Professor at Temple University in 2010 and the University of Colorado Denver in 2011. Notable achievements include serving as the Director of the Science and Mathematics Teaching Center at UW (2012–2016) and as Fulbright Canada Research Chair in STEM Education at the University of Calgary (2018). Her research interests are culturally specific pedagogy, teaching mathematics for social justice, and computational thinking/participation in STEM.

Jakita O. Thomas, PhD, is the Philpott Westpoint Stevens Associate Professor of Computer Science and Software Engineering at Auburn University in Auburn, AL, United States. Dr. Thomas is also Director of the **CU**ltu**R**ally & **SO**cially **R**elevent (**CURSOR**) Computing Lab at Auburn. Prior to her current position at Auburn, Dr. Thomas was a member of the Computer & Information Science faculty at Spelman College (2010–2016), where she was awarded the Spelman College Presidential Award for Excellence in Teaching by a Junior Faculty Member (2013) as well as the Spelman College Presidential Award for Scholarly Achievement by a Junior Faculty Member (2015). Dr. Thomas is also a recipient of the National Science Foundation's Faculty Early Career Development (CAREER) Award (2012–2019). She is also a recipient of the Presidential Early Career Award for Scientists and Engineers (PECASE) (2016).

Roni Ellington, PhD, is Associate Professor of Mathematics Education at Morgan State University in Baltimore, MD, United States. Dr. Ellington is the former coordinator of Graduate Programs in Mathematics and Science Education at Morgan State University. She was awarded the Morgan State University Teaching

Excellence Award in 2014, and she was a featured speaker at the inaugural Baltimore TEDx Conference in 2013. Her research interests include understanding the experiences of high-achieving African-American students in mathematics/STEM, interdisciplinary STEM education, professional development of K–12 STEM teachers, and equity and diversity in undergraduate STEM education.

Monica B. Mitchell, EdD, is Founder and President of MER Associates, LLC (MERA). Recognized as a top 100 minority business enterprise in the greater Washington, DC, area, MERA has a particular focus on efforts that broaden science, technology, engineering, and mathematics (STEM) participation. Prior to MERA, Dr. Mitchell served as a Program Officer at the National Science Foundation (2002–2005), and New Visions for Public Schools (1996–2001). She is also an Affiliate Faculty member and Researcher at the Center for Culturally Responsive Evaluation and Assessment (CREA), University of Illinois Urbana-Champaign.

Olatokunbo (Toks) S. Fashola, PhD, is Research Professor at American University, Washington, DC, United States. Dr. Fashola is currently the Faculty Coordinator for American University's Dual Enrollment program, and she is the Principal Investigator for two CTE grants with District of Columbia Public Schools. Prior to accepting the position at American University, Dr. Fashola also served as a Research Scientist and a Faculty Associate at the Johns Hopkins University. She is also a Principal Research Scientist and Vice President for Evaluation at MER Associates, LLC (MERA). At MERA, Dr. Fashola evaluates several projects that address STEM among underrepresented minority undergraduates.

1

THE ADVENT OF COMPUTATIONAL THINKING

Jacqueline Leonard and Jakita O. Thomas

Six children built a go-cart using the wheels and frame of a little red wagon, wood, rope, and cardboard boxes. They nailed wood horizontally above and beneath the frame to make the chassis. Another piece of wood was nailed vertically at a 90-degree angle across the chassis to turn the go-cart right or left with their feet. A rope was also tied to the frame to steer wheels, and the sides of the go-cart were made of cardboard boxes. The go-cart was placed at the top of the entrance to an alleyway, and the children took turns riding. When it was Jenni's (pseudonym) turn, she found it difficult to steer the go-cart because it was traveling faster than she anticipated. Jenni turned sharply to the left and crashed. With minor injuries, she and the other children took the go-cart to the yard and tinkered with it to make the steering more manageable. These children, all under the age of 12, were limited only by their imagination and the materials they had on hand. This event could have taken place in any neighborhood regardless of place, ethnicity, or race. However, the go-cart was created by Black children during the 1960s. They were engaged in STEM before the term was ever spoken. They used engineering to design the go-cart, physics when they placed it at the top of the alleyway, mathematics when they placed horizontal and vertical parts at 90-degree angles, and computational thinking (i.e., conditional reasoning) to improve the steering. Moreover, they engaged in computational participation as evidenced by their planning, trial runs, and tinkering to improve the go-cart.

The vignette mentioned earlier is an authentic example of computational thinking (CT), which was popularized and defined by Wing (2006) as a "problem-solving approach that draws on concepts fundamental to computer science by 'reformulating a seemingly difficult problem into the one we know how to solve, perhaps by reduction, embedding, transformation, or simulation'" (p. 33). The children in the vignette used reduction to solve the go-cart problem by examining how each part functioned in concert with the others. Consistent with Wing's (2006) argument that CT is not just for computer scientists, these children were engaged in algorithmic thinking approaches similar to those posited by Pólya (1973), Sengupta et al. (2013), and others regarding the problem-solving process (i.e., translating, integrating, planning, and executing).

The impetus for building the go-cart, in part, was the popular television show, *Speed Racer* (Yoshida, 1967–1968) in which Trixie was a female character. Thus, the children translated what they saw in the cartoon into an actual image. Integration involved finding the necessary parts to create the go-cart. Planning occurred during the initial construction and refinement stages as the children tinkered with the go-cart to optimize its performance. Execution took place when the children tested the go-cart by steering it down the alleyway. Such occasions provide children with authentic experiences that predispose them to careers in science, technology, engineering, and mathematics (STEM). To illustrate this point, Jenni (pseudonym) in the aforementioned vignette is the first author of this text, Jacqueline Leonard. She recalled the excitement that she felt while building the go-cart and realized it was an example of CT. As a Black woman, Jacqueline believes she is fortunate to be born after *Brown vs. Board of Education* (1954). What factors influenced her to become a mathematics educator/researcher?

Recollections of her childhood and schooling revealed a broad interest in mathematics and science. Jacqueline participated in at least two science fairs while attending a predominantly Black urban K-8 public school, and school records showed a composite Iowa Test of Basic Skills score of 11.0 (i.e., 11th grade, 0-month grade-level equivalency) in grade eight. Mathematics was her favorite subject, and she excelled in algebra I, algebra II/trigonometry, biology, and chemistry in high school. Obtaining a full scholarship to Boston University she double-majored in physical therapy and pre-med. Although she worked as a candy striper at a hospital in her community where she was mentored by a medical supply technician, Jacqueline did not have a clear understanding of what physical therapy entailed. Lacking exposure to role models and being unfamiliar with health science and medicine, Jacqueline decided to become a science and mathematics teacher. Black women in her community, as well as an aunt, were teachers and role models. Moreover, she enjoyed sharing her knowledge and enthusiasm with children to ensure they had adequate preparation and opportunities to succeed. Jacqueline completed a Bachelor of Arts degree and teacher certification at Saint Louis University. Her love of learning later led her to obtain a Master of Arts in Teaching Mathematical Sciences from the University of Texas at Dallas. A teacher colleague—Dr. William F. Tate IV—who obtained his PhD at the University of Maryland and became an assistant professor in Wisconsin, led her to do the same. In 1997, Jacqueline received a PhD in Curriculum and

Instruction and began a tenure-track position in mathematics education at Temple University.

Imagine what Black[1] children in the 21st century may invent or create using Shuri from the movie, *Black Panther* (Coogler, 2018), as their role model or impetus. What gendered and racial examples do children see and experience in the 21st century that allow them to see themselves in STEM? What impact will the COVID-19 pandemic have on children's interest in STEM in this decade? Understanding how viruses grow exponentially and how computers can be used to model growth curves are of interest to society at large as well as K-12 students. The foregoing vignette, as well as others presented in this text, will be used as a springboard for discussing research and practice that foster CT and computational participation among underrepresented students, including girls and students of color. Strategies that have been successful in broadening STEM participation for underrepresented and underserved K-12 students are advanced in this work, which begins with self-efficacy and expectancy value as conceptual frameworks, which are followed by the underpinnings of CT.

Theoretical Framework

The projects referenced in this text were funded by the National Science Foundation's (NSF), Innovative Technology Experiences for Students and Teachers (ITEST) program. Two projects—uGame-iCompute (2013–2017) and Bessie Coleman (2018–2021)—were grounded in self-efficacy theory (Bandura, 1997) and expectancy-value theory (Wigfield & Eccles, 2000). Bandura (1997) contended that mastery experiences, vicarious experiences, verbal persuasion, and affective states contribute to efficacy. Mastery experiences include activities that allow a person to develop skills in a particular field, such as serving as an intern in a specific occupation. Schunk (2020) contended that observing or listening to human, nonhuman (e.g., cartoon characters or animated creatures), or multimedia sources (e.g., television, movies, DVDs, books, etc.) results in vicarious learning. Research has shown that role models (Bracey, 2013; Jacob et al., 2018; Ryoo, 2019; Thomas et al., 2018), whether real or imaginary, are important vicarious figures when it comes to broadening participation in STEM. Efficacy drawn from vicarious experiences entails seeing the qualities of others who excelled at a particular endeavor in oneself. Verbal persuasion occurs when a perceived expert affirms or disaffirms one's ability to pursue a specific career. Thus, parents, teachers, and counselors can play an important role in students' persistence in STEM. Affective states refer to students' preferences or desire to pursue a specific career. Exposure to specific career paths may lead students to pursue those paths. One purpose of the projects was to expose underrepresented students to computer science through robotics/game design (i.e., uGame-iCompute) and computer modeling/drones (i.e., Bessie Coleman Project).

In terms of expectancy-value theory, Wigfield and Eccles (2000) contended that "individuals' choice, persistence, and performance can be explained by their beliefs about how well they do on [an] activity and the extent to which they value the activity" (p. 68). The value one places on a subject or task predicts future participation in similar subjects or tasks (e.g., enrolling in advanced STEM courses, pursuing STEM careers, etc.). Leonard (2019), Ryoo (2019), Scott et al. (2015), and Thomas et al. (2018) are some of the researchers who suggest that culturally responsive computing and intersectional computing are needed to increase access and equity in computer science. Thus, role models, culturally responsive computing, and intersectional computing along with educational innovation and real-world applications, are crucial elements for sustaining interest, increasing self-efficacy, and developing STEM identity among females and ethnically/racially diverse students.

The Emergence of Computational Thinking in Education

Yadav et al. (2017) traced the development of CT from the problem-solving process work of Pólya (1973), a mathematician whose *How to Solve It* publication made important contributions to mathematics education. Without the benefit of computer science, Pólya articulated how to solve problems in a systematic way (Yadav et al., 2017). In the 1980s, CT became associated with Seymour Papert, who pioneered the notion that children can use procedures to engage in LOGO programming (Grover & Pea, 2013). In Papert's seminal text entitled, *Mindstorms* (1980, 1993), the LOGO programming language provided an example of how technology can be used to promote CT (Kafai & Burke, 2014; Leonard, 2019). Papert believed that LOGO provided students with opportunities to use the computer to influence how people think (Yadav et al., 2017). This is important because ability to think and engage in mental processes is the crux of learning and cognition (Schunk, 2020).

In the 1950s and 1960s, problem solving gave way to algorithmic thinking (Rich & Langston, 2016), which is defined as the ability to formulate a solution to a problem in algorithmic form and then implement it as a computer program (Syslo, 2015). Yet, others emphasize the importance of using a computer to solve problems (Barr & Stephenson, 2011). Moreover, Sengupta et al. (2013) contend that CT allows students to deal with abstractions, which allow children to receive inputs that lead to executing a series of steps that produce outputs to satisfy a specific goal (Wing, 2006). However, CT goes beyond technology and computer science and is applicable to a variety of activities including computer modeling, robotics, and game design (Leonard et al., 2016, 2018; Sengupta et al., 2013; Sullivan & Heffernan, 2016).

Defining Computational Thinking

Wing (2006) was the first to propose a general definition of computational thinking, which she described as "solving problems, designing systems, and understanding human behavior by drawing on the concepts fundamental to computer science" (p. 33). She later defined computational thinking as "the thought processes involved in formulating problems and solutions so that the solutions are represented in a form that can be effectively carried out by an information-processing agent" (Wing, 2010, p. 1). With many scholars and educators beginning to recognize CT as a competence domain (Yadav et al., 2017), several other definitions of CT have emerged (Grover & Pea, 2013; Wing, 2008). One of the simplest definitions describes CT as formulating problems in such a manner that they can be solved by "computational steps and algorithms" (Aho, 2010, as cited in Grover & Pea, 2013). While these competencies can be applied to any subject area (Chen et al., 2017; Yadav et al., 2017), CT and programming have yet to be fully integrated with K-12 curricula, including science where CT can be incorporated to help students develop scientific reasoning (Sengupta et al., 2013).

The International Society for Technology in Education (ISTE, 2011) standards provide a comprehensive list of dimensions that can be used across disciplines as evidence to assess students' CT in different content domains (Leonard et al., 2016; Newton et al., 2020). According to Grover and Pea (2013), the dimensions that are widely accepted to form the basis for curriculum and assessment include the following: (a) abstractions and pattern generalizations; (b) systematic processing of information; (c) symbol systems and representations; (d) algorithmic notions of flow and control; (e) iterative, recursive, and parallel thinking; (f) conditional logic; (g) efficiency and performance constraints; and (h) debugging and systematic error detection. Thus, CT, in the U.S. and other parts of the world, has been described as a set of competencies that involve students' development of "domain specific and general problem-solving skills" (Yadav et al., 2017, p. 49).

Computing and research-focused organizations (e.g., National Research Council [NRC], Association of Computing Machinery [ACM], NSF, Computer Science Teachers of America [CSTA] and others) have outlined practices that are essential to K-12 science, engineering, and computer science education (CSTA, 2018a; NRC, 2011; K-12 Computer Science Framework, 2016). Among those practices are: (a) defining problems; (b) developing and using models (including prototypes); (c) analyzing and interpreting data; (d) designing solutions; (e) engaging in argument from evidence; (f) obtaining, evaluating, and communicating information; (g) fostering an inclusive computing culture; (h) collaborating around computing; (i) recognizing and defining computational problems; (j) developing and using abstractions; (k) creating computational artifacts, testing, and refining computational artifacts; and (l) communicating about computing (CSTA, 2018b). The K-12 Computer Science Framework (2016) further states that CT "plays a key role in the computer science practices of the framework as it encompasses" recognizing and defining computational problems, developing and using abstractions, creating computational artifacts, and testing and refining computational artifacts (p. 52). As learners acquire and develop CT skills and capabilities, they should have opportunities to "plan and carry out" design projects in which they "define problems in terms of criteria and constraints, research the problem to deepen their relevant knowledge, generate and test possible solutions, and refine their solutions through redesign" (NRC, 2011, p. 71).

Furthermore, the development of CT assessments is underway in formal and informal PreK-12 settings. Much of the work in CT assessment development has been supported through the NSF, inclusive of programs in the Directorate for Education and Human Resources (EHR) (e.g., the Discovery Research PreK-12 program) and the Computer and Information Science and Engineering (CISE) Directorate (e.g., the STEM + Computing K-12 Education program). While some of the formal CT assessments under development are tied directly to specific curricula, others have broader integrative and interdisciplinary foci. As more states establish policies promoting computing and computer science education across the K-12 continuum (CSTA, 2018a), the development and availability of curricula and formal assessments to support CT learning at scale will follow. In the interim, recent educational interventions as well as research and development projects that address the teaching and learning of CT shed light on promising strategies to inform current instruction and assessment for all students. Our focus throughout each chapter centers on opportunities to further CT teaching and learning among underrepresented and underserved students of color.

Promoting Computational Thinking Among Elementary School Students

Economic factors and career readiness programs are driving forces for broader integration of computer science in K-12 education (Chen et al., 2017). However, the research is sparse when it comes to assessing CT among elementary students (Chen et al., 2017), especially those from underrepresented backgrounds (Newton et al., 2020). Tools that meet the established criteria of providing a low floor (i.e., basic entry), high ceiling (i.e., advanced learning), and wide walls (i.e., inclusion of experience and culture) (Kafai & Burke, 2014) to foster CT among elementary-aged students include Scratch, Alice, Game Maker, Kodu, AgentSheets, AgentCubes, Arduino boards, Makerspaces, and robotic kits (Grover & Pea, 2013). In this text, we examine CT using a variety of tools (e.g., LEGO® robotics kits, AgentSheets, AgentCubes, Tinkercad, Flight Simulator X, Makerspaces, and UAVs) to allow for scaffolding, learning transfer, equitable practices, and sustainability (Grover & Pea, 2013; Repenning et al., 2010) in urban, suburban, and rural contexts.

Chen et al. (2017) conducted a study to develop an instrument to assess elementary students' CT using text-based (i.e., word problems) and visual programming (i.e., robotics). The researchers reported an Expected A-Posteriori/Plausible Value (EAP/PV) reliability of 0.81, which fell within an acceptable range. Drawing on the CSTA (2018b) standards, Chen et al. developed a five-component assessment framework to assess 767 students at a single school (Gender: 48.8% female; 51.2% male; Race/Ethnicity: 40.8% Black, 24.8% Hispanic, 23.1% White, 6.9% Asian, and 4.4% Other). The test consisted of 15 multiple choice and 8 open-ended items for a total of 23 items. A three-point rubric was created to score the open-ended items. Interrater reliability for the rubric was 85%. The T-statistic was used to determine if there were significant differences among fifth graders who were assessed using two forms of the test (text-based vs. visual programming). The text-based items could be described as traditional word problems, and the visual programming items involved using a humanoid robot to perform functions such as communication, movement, or dance. The results of a two-tailed, two sample t-test found no significant differences between groups by type of test. However, when six classes (n=125) were compared, results showed one class outperformed all of the others. When the lowest performing class (Class I) and the highest performing class (Class II) were compared and contrasted, the results revealed that students in both groups had higher gains in the robotics programming context [Class I ($t = -3.562; p = 0.001; d = 0.76$) & Class II ($t = -2.901; p = 0.0005; d = 0.73$)] than the traditional context [Class I ($t = -1.560; p = 0.067; d = 0.33$) & Class II ($t = -2.746; p = 0.008; d = 0.69$)]. Chen et al.'s study is interesting from a curriculum standpoint. Students in Class I did better on the context specific robotics curriculum than they did on traditional word problems. Likewise, they received higher ratings on open-ended items on the visual programming test, which allowed them to show their CT in robotics by drawing or writing out the programming steps. This finding concurs with prior research that suggests context specific computer-based programming improved students' problem-solving skills (Kinard & Bitter, 1997; Leonard et al., 2005).

Leonard et al. (2016) conducted a study on robotics and game design called uGame-iCompute that was implemented from 2013 to 2017 in Wyoming and Pennsylvania. Students used LEGO® NXT and EV3 robotics to perform tasks such as playing a song, moving an object, and following an obstacle course. MINDSTORMS® programming was used to get the robots to perform the given tasks. Students also created games using AgentSheets and AgentCubes (Repenning et al., 2010). A rubric was used to score students' CT during game design. Interrater reliability was established at 86%.

During the pilot study, which took place in Wyoming, the researchers isolated robotics from game design to determine whether there were significant differences on subscales that measured rural students' self-efficacy in videogaming, computer gaming, and using the computer to solve problems. In order to do this, the researchers implemented a single treatment (i.e., robotics only or game design only) in three classrooms and two treatments (i.e., robotics/game design) in five classrooms (Leonard et al., 2016). The results of the pilot revealed that combining robotics and game design resulted in higher scores (see Table 1.1) In Year 3, the study was expanded to include urban students in Pennsylvania (Leonard, 2019). The results of a paired t-test revealed that two treatments (i.e., robotics/game design) were more robust when the self-efficacy scores of students in Wyoming and Pennsylvania were combined. As a result, students showed significant gains on the constructs of videogaming, computer gaming, and computer use (see Table 1.2). This finding indicates that increasing the sample size improved statistical power.

TABLE 1.1 Mean SETS Post-Survey Scores by Type of Instruction (Year 1)

Group (n=49)	Number of Students	Pre-Survey Mean	Standard Deviation	Post-Survey Mean	Standard Deviation
Videogaming					
Group 1 (G)	29	3.79	0.92	3.75	0.93
Group 2 (R/G)	20	3.68	0.62	4.14★	0.82
Computer Gaming					
Group 1 (G)	29	3.57	0.77	3.43	0.88
Group 2 (R/G)	20	4.02	0.73	4.02	0.73
Computer Use					
Group 1 (G)	29	4.06	0.71	3.61	0.88
Group 2 (R/G)	20	4.17	0.64	4.32	0.68

★ $p < 0.05$

TABLE 1.2 Analysis of Elementary Students' SETS Survey (Year 3)

Constructs	Pretest	Standard Deviation	Posttest	Standard Deviation	Gain Score
Videogaming					
All students (n=197)	4.04	0.67	4.14★	0.65	0.10
Wyoming (n=137)	4.04	0.68	4.14	0.65	0.10
Pennsylvania (n=60)	4.04	0.64	4.14	0.66	0.10
Computer Gaming					
All students (n=196)	3.95	0.69	4.09★	0.65	0.14
Wyoming (n=136)	3.92	0.71	4.08★	0.67	0.16
Pennsylvania (n=60)	4.04	0.65	4.11	0.61	0.07
Computer Use					
All students (n=195)	3.92	0.82	4.04★	0.74	0.12
Wyoming (n=135)	3.87	0.83	4.02★	0.75	0.15
Pennsylvania (n=60)	4.02	0.80	4.09	0.73	0.07

★ $p < 0.05$

Computational Thinking: Language, Gender, and Race/Ethnicity

While research on CT is sparse at the elementary school level, studies that focus on language, gender, and race/ethnicity are even rarer. In the midst of increases in funding initiatives for girls and underrepresented students during the last decade (Ryoo, 2019), gender and racial diversity in computing continues to be dismal. Furthermore, aging school buildings, inequitable school practices, high teacher and administrator turnover, and negative teacher perceptions about poor students and students of color have a profound impact on who is exposed to and who excels in computer science (Leonard, 2019; Ryoo, 2019). Recall that Jacqueline (first author of this text) abandoned the pursuit of a career in the health sciences or medicine because she did not have any role models or familiarity with the fields. Children's self-efficacy has been shown to influence the types of careers in which they perceive their own potential success or failure (Bandura et al., 2001, as cited in Schunk, 2020). Knowing that others from her background succeeded in mathematics, a subject that she also enjoyed, increased her self-efficacy and convinced Jacqueline that pursuing an advanced degree in mathematics education was an attainable goal.

Computational Thinking among English Learners

Coding provides a computing environment for English learners (ELs) to develop problem-solving skills, academic language proficiency, and content knowledge (Jacob et al., 2018). The CONECTAR (Collaborative Network of Educators for Computational Thinking for All Research) researchers developed and piloted instructional materials to teach CT to Hispanic students in California (Jacob et al., 2018). The goal of the CONECTAR Project was to examine CT among third- through fifth-grade ELs (60%) and Latinx students (96%) in the Santa Ana Unified School District. The Scratch programming language was used by five teachers to develop approaches to improving academic language and developing computational fluency among ELs. Students received explicit vocabulary instruction, with computer science and language objectives, to complement their learning of basic Scratch elements. Free choice and the open-ended nature of the Scratch projects allowed students to incorporate elements of their identity into the finished products. Teachers were also able to scaffold various levels of programming and language competence. Thus, the curriculum was flexible and culturally relevant, drawing on storybooks and role models, such as prominent female scientists, to motivate students' interest-driven designs and digital games.

In Project STITCH (Jacob et al., 2018), 18 teachers implemented a project that engaged Latinx students in grades 6–12 in e-textiles. Middle and high school students used "microprocessors to gather and process data needed to solve a range of authentic problems drawn from physics, chemistry, earth science, and life science" (Jacob et al., 2018, p. 19). Some examples of e-textiles are heated jackets and light-up footballs (Barton & Tan, 2018). Arduinos can be also used to create purposeful e-textiles, which allow ELs to engage in abstraction, logical thinking, generalization, and debugging (i.e., CT) while also increasing their vocabulary and academic language proficiency. These authentic products allowed underrepresented youth to draw on community assets, sense making, and agency to create products they believed were useful and could be beneficial to others (Barton & Tan, 2018; Jacob et al., 2018).

Project STITCH engaged students in "culturally relevant learning ... through a language-neutral set of projects [that] allowed for science learning" (Jacob et al., 2018, p. 20). Perhaps one of the most important outcomes of the project, however, was teacher change. As a result of working with students on e-textiles, teachers' beliefs about their students and their ability to *do* science changed. Shifting teacher perspectives as well as engaging ELs and other underrepresented students in authentic tasks can influence their self-efficacy, expectancy value, and motivation to foster a new generation of computer scientists.

Computational Thinking at the Intersections among Black Girls

Computational algorithmic thinking (CAT), a scaffolded on-ramp to CT, is defined as "the ability to design, implement, and assess the implementation of algorithms to solve a range of problems" (Thomas et al., 2015, p. 16). CAT focuses on the algorithms designed, adapted, implemented, and discarded by the human (as computing agent) on the journey toward choosing the "right" abstractions (Wing 2008; Thomas et al., 2016). CAT ...

> involves identifying and understanding a problem, articulating an algorithm or set of algorithms in the form of a solution to the problem, implementing that solution in such a way that it solves the problem, and evaluating the solution based on some set of criteria.
>
> (Thomas et al., 2015, p. 16)

Supporting Computational Algorithmic Thinking (SCAT) is a six-year longitudinal study that aimed to not only explore the development of CAT capabilities in Black girls as they worked in dyads and developed a set of games for social change but also to understand the impact of a program like SCAT on their perceptions of themselves as game designers and critical thinkers. Participants, called SCAT Scholars, engaged in three types of activities each year: (a) a two-week summer design experience; (b) twelve academic-year workshops where dyads iteratively refined their games and submitted them to local and national game design competitions using visual programming languages and game engines (e.g., Scratch, App Inventor, & Unity); and (c) field trips where Scholars could experience the application of CAT in different contexts and for different career pathways (e.g., Georgia Tech Aware Home, Museum of Design Atlanta, Spelman College Innovation Lab, etc.). SCAT has engaged a cohort of 23 Black girls who started the program the summer before their sixth-grade year and will complete the SCAT program as graduating high school seniors in 2020.

SCAT created a safe computing space for scholars to find meaning in relevance and altruism as they designed games that addressed social and culturally relevant issues in their communities to author creative imaginations in the production of knowledge and games, and to create new narratives about themselves and other Black women and girls, thereby disrupting deficit models (Joseph & Thomas, 2020). The intentional centering on Black girls and women in SCAT and their SCAT experiences overall impacted Scholars' perceptions of themselves as game designers, and therefore, critical and computational thinkers, with the percentage of Scholars perceiving themselves as game designers shifting from 0% to almost 50% over the course of 3 years (Joseph & Thomas, 2020; Thomas et al., 2016). This shift not only suggests that learners need time (much more than a semester or a year) to reflect on their computer science experiences in such a way that they can progress from *questioning* who does computer science (e.g., "Black women don't do CS." or "Do Black women do CS?") to *stating* who does do computer science (e.g., "Black women DO CS.") to, ultimately, *personalizing* who does computer sciences (e.g., "I can do CS."). Such leaps are non-trivial. Learners traditionally at the margins of computing also need spaces to experience computer science education that center, care about, and celebrate them, honoring the expertise and funds of knowledge they bring to these spaces, while affirming who they are through the artifacts they create and imagining who they can become by showcasing people at different places along the computing trajectory who look like them.

Computational Thinking among Urban Students

The 3-year uGame-iCompute study cited earlier also examined outcomes among predominantly Black student participants who resided in a large urban city (Newton et al., 2020). Demographics revealed the sample was 85% Black/African American, 15% White/Caucasian, and 60% male and 40% female. Approximately 113 urban students in grades three to six engaged in LEGO® robotics and game design through the use of a program called AgentSheets (Repenning et al., 2015). When SETS scores were combined for two different cohorts in Pennsylvania (afterschool and summer camp), the self-efficacy scores of predominantly Black students significantly increased on computer gaming (see Table 1.3). However, scores on videogaming remained stable while scores on using the computer to solve problems (i.e., CT) declined. Attitudes toward engineering and technology remained stable. Yet, it is possible that the survey results were influenced by ceiling effects or the instrument did not fully capture CT since videogaming and computer gaming are not the same as programming robots and designing digital games.

TABLE 1.3 Results of Urban Elementary Students' SETS and Engineering/Technology Attitudes

Construct	Pre-Survey M (SD)	Post-Survey M (SD)	t
Videogaming ($n = 93$)	4.07 (0.67)	4.09 (0.73)	−0.502
Computer Gaming ($n = 91$)	4.02 (0.68)	4.16 (0.60)	−1.994★
Using Computer ($n = 91$)	4.10 (0.74)	3.97 (0.69)	0.040
Engineering/Technology ($n = 92$)	4.14 (0.71)	4.14 (0.74)	1.457

★ $p<0.05$

The research team did not have an assessment tool to assess both robotics and game design as teachers used different strategies to engage students in CT in these two contexts. Thus, the Sullivan and Heffernan (2016) learning progression model was used to assess CT in robotics, and an earlier rubric (Leonard et al., 2016) was modified to use Repenning et al.'s framework (2015) and the ISTE standards (2011) to assess CT in game design (Newton et al., 2020). Qualitative data revealed robust findings. Predominantly, Black urban focal students exhibited substantive CT during robotics as measured by a learning progression model and moderate CT during game design as measured by the revised rubric. Student interviews during focus groups were transcribed and analyzed to triangulate the data. Evidence of students' CT in robotics and game design are presented subsequently. Pseudonyms are used for anonymity.

> Actually, I had to connect it to a computer and then I had to put all these different codes in, but that way that one [LEGO]…was, it was different.
>
> (Cameka)

> I liked that we could do it with groups so we could get everyone's idea about how the robot would be built, and what I liked about the computer is the way…how we set them up.
>
> (Kenadi)

> I liked the way how like we would get to design the um robots because… like a character or like a car and like the bus ride, too, and like it was fun and we could sing songs and…program the games.
>
> (Aedan)

> I found out that you could make a game like directly on the computer. Like I made a game where I had to draw it and then just repeating it on the computer but…. Now, I know that I can make it directly.
>
> (Tara)

I used the computer to solve problems by like if I couldn't draw something, for my AgentSheets, I'd go online, and look up pixelated characters.

(Shelly)

There's a glitch going on right now, I was trying to help my friend.... I don't know if it's her computer or AgentSheets.... Her sound is not going through all the way.

(Mitch)

...me and my partners were programming our character to be like swooshed when the car ran over her, but every time we put her in front of the car she didn't get squished.... She just stopped the cars, [and] she just blocked...the cars. That drives me nuts! And we never got [AgentSheets] to work.

(Zoe)

Analyses of the focus group data show that students made explicit and implicit inferences to CT: *coding* (2 instances), *computing* (6 instances), *program(ming)* (2 instances), *design* (3 instances), *group work* (3 instances), *glitches* (2 instances), and *problem solving* (2 instances). All of these terms are associated with CT. Students described coding and programming during robotics and game design in terms of abstractions and generalizations. Algorithmic flow and logical thinking were also described when Tara mentioned finding a shortcut to programming in game design. Shelly provided an example of iterative, recursive, and parallel thinking when she described the ability to draw using pixels. Mitch and Zoe mentioned a problem with glitches. Programming glitches allow students to engage in debugging, which is also evidence of CT. While it is unclear if Mitch was able to fix the glitch, Zoe reported they never got their program to work. Frustration with debugging could explain the decline on the SETS subscale—using the computer to solve problems—in the Pennsylvania setting. Nevertheless, group work is a hallmark of computational participation (Kafai & Burke, 2014).

Overall, the results of the Year-3 uGame-iCompute study were mixed. Quantitative data showed students' self-efficacy on computer games increased significantly, while videogaming and use of the computer constructs did not change significantly. Attitudes toward engineering/technology were unchanged while qualitative data showed students exhibited moderate and substantive CT.

Summary

The research studies cited earlier suggest that elementary students' engagement in various activities can contribute to their development of CT. Students should not only take control of their own learning in such settings but inspire their peers as well. Prior research shows that engagement in communal rather than individual work not only promotes the social dimension of coding (Kafai & Burke, 2014) but has the potential to increase student achievement as well (Coleman et al., 2016). Developing underrepresented students' CT must go beyond workforce preparation (i.e., interest convergence) to foster civic engagement that allows children to use their innovation to mitigate social ills (Jacob et al., 2018). For example, during the COVID-19 pandemic, students in Soacha, Colombia, made face masks to protect themselves and their community from the coronavirus (Barth, 2020). In the U.S., students in Charlotte, North Carolina, raised funds to print 3D face shields for medical personnel to use during the outbreak (WBTV Web Staff, 2020). Creating meaningful and life-saving devices transforms CT to computational participation. Thus, emerging technology and Makerspaces should be used to allow children to participate in universal design and engage in computational participation (Kafai & Burke, 2014).

The studies cited earlier also point to the need to design instruments and rubrics that more accurately reflect the CT strategies that young learners engage in and promote CT skills that are harder to isolate and measure, such as systematic processing of information (i.e., data processing) (Chen et al., 2017). As research in CT expands, the need to share rubrics and assessments that are effective with K-12 students also increases. Without proper assessment, CT cannot be sustained as a vital part of K-12 curriculum (Grover & Pea, 2013). The purpose of this text is to not only share such strategies but to ensure that all students, including girls and underrepresented minority students, engage in authentic curricula that promote CT writ large and broaden participation in STEM.

As shown in the opening vignette, Black children are capable of engaging in CT if they are provided with unique opportunities to do so. In the vignette, the learning context was organic, and the children, inspired by cartoon characters, decided to engage in an engineering activity on their own. Recall that nonhuman characters are also a source of vicarious learning (Schunk, 2020). Nevertheless, there is a role for teachers and mentors to play in promoting racially equitable computing. Recent studies of self-efficacy and motivation among ELs (Jacob et al., 2018) and predominantly Black students (Leonard et al., 2016) reveal that authentic activities such as Scratch programming, LEGO robotics, and game design are promising.

The results of the aforementioned studies reveal there is more work to be done to improve the quality of computer science education in the United States. Introducing K-12 students to computer science through coding and computer programming is currently in the forefront of educational reform with some countries developing content specific and/or general technology courses (Falloon, 2016; Rich & Langston, 2016; Yadav et al., 2017). As school-based access to technology increases, educational uses of technology are also expanding throughout the curriculum in formal and informal learning environments. With this expansion, there is a critical need to identify the tools, pedagogy, and practices deemed essential to promote CT, especially given the data-rich context in which we currently live.

This chapter provided an overview, prevailing definitions of CT and CAT, and examples of recent studies to address CT in K-8 schools, and a description of how CT/CAT have impacted diverse learners. The remaining chapters will delve into these topics in more detail to provide deeper understandings of CAT and coding (Chapter 2) and innovative curriculum, specifically using robotics, digital games, computer modeling, and drones as the impetus to promote CT (Chapters 3–5). Finally, we address professional development and best practices for modeling and integrating culturally responsive CT (Chapter 6) and culturally responsive evaluation (Chapter 7) to round out fostering CT among underrepresented students for the purpose of supporting racially equitable computing.

Note

1 Black is used as an inclusive term to describe all peoples from the African diaspora.

References

Bandura, A. (1997). *Self-efficacy: The exercise of control.* W. H. Freeman.

Barr, V., & Stephenson, C. (2011). Bringing computational thinking to K-12: What is involved and what is the role of the computer science education community? *ACM Inroads, 2*(1), 48–54. https://doi.org/10.1145/1929887.1929905

Barth, D. (2020, March 25). There's a global shortage of face masks—but people around the world are using creative methods to make their own. *Business Insider.* https://www.businessinsider.com/coronavirus-make-own-face-mask-2020-3

Barton, A. C., & Tan, E. (2018). *STEM-rich maker learning.* Teachers College Press.

Bracey, J. (2013). The culture of learning environments: Black student engagement and cognition in math. In J. Leonard, & D. B. Martin (Eds.), *The brilliance of black children in mathematics: Beyond the numbers and toward new discourse* (pp. 171–194). Information Age Publishing.

Brown v. Board of Education, 347 U.S. 483 (1954).

Chen, G., Shen, J., Barth-Cohen, L., Jiang, S., Huang, X., & Eltoukhy, M. (2017). Assessing elementary students' computational thinking in everyday reasoning and robotics programming. *Computers & Education, 109,* 162–175. https://doi.org/10.1016/j.compedu.2017.03.001

Coenraad, M., Ketelhut, D. J., Leonard, J., & Jordan, W. J. (2020, April 17–21) *The Computational Thinking Self-Efficacy (CompTSE) Scale validation* [Roundtable Session]. Annual Meeting, of the American Educational Research Association, San Francisco, CA, United States. http://tinyurl.com/ww9poaz (Conference Canceled).

Coleman, S. T., Bruce, A. W., White, L. J., Boykin, A. W., & Tyler, K. (2016). Communal and individual learning contexts as they relate to mathematics achievement under simulated classroom conditions. *Journal of Black Psychology, 43*(6), 543–564. https://doi.org/101177/0095798416665966

Computer Science Teachers Association [CSTA]. (2018a). *State of computer science education: Policy and implementation.* https://advocacy.code.org/

Computer Science Teachers Association [CSTA]. (2018b). *Computer Science (CS) standards.* https://www.csteachers.org/page/standards

Coogler, R. (Director). (2018). *Black panther* [Film]. Marvel Studios.

Falloon, G. (2016). An analysis of young students' thinking when completing basic coding tasks using Scratch Jnr. on the iPad. *Journal of Computer Assisted Learning, 32*(6), 576–593. https://doi.org/10.1111/jcal.12155

Grover, S., & Pea, R. (2013). Computational thinking in K-12: A review of the state of the field. *Educational Researcher, 42*(1), 38–43.

ISTE. (2011). *Operational definition of computational thinking for K-12 education.* https://id.iste.org/docs/ct-documents/computational-thinking-operational-definition-flyer.pdf

Jacob, S., Nguyen, H., Tofel-Grehl, C., Richardson, D., & Warschauer, M. (2018). Teaching computational thinking to English learners. *NYS TESOL Journal, 5*(2), 13–24.

Joseph, N. M., & Thomas, J. O. (2020). *Designing STEM learning environments to support middle school Black girls' computational algorithmic thinking: A possibility model for disrupting STEM neoliberal projects* [Paper presentation]. International Conference of the Learning Science, Nashville, TN, United States. (Conference canceled).

K-12 Computer Science Framework. (2016). https://k12cs.org

Kafai, Y. B., & Burke, Q. (2014). *Connected code: Why children need to learn programming.* The MIT Press.

Kinard, B., & Bitter, G. G. (1997). Multicultural mathematics and technology: The Hispanic Math Project. *Computers in the Schools, 13*(1/2), 77–88.

Leonard, J. (2019). *Culturally specific pedagogy in the mathematics classroom: Strategies for teachers and students* (2nd ed.). Routledge.

Leonard, J., Barnes-Johnson, J., & Evans, B. R. (2019). Using computer simulations and culturally responsive instruction to broaden urban student's participation in STEM. *Digital Experiences in Mathematics Education, 5,* 101–123. https://doi.org/10.1007/s40751-018-0048-1

Leonard, J., Buss, A., Gamboa, R., Mitchell, M., Fashola, O. S., Hubert, T., & Almughyirah, S. (2016). Using robotics and game design to enhance children's STEM attitudes and computational thinking skills. *Journal of Science Education and Technology, 28*(6), 860–876. https://doi.org/10.1007/s10956-016-9628-2

Leonard, J., Davis, J. E., & Sidler, J. L. (2005). Cultural relevance and computer-assisted instruction. *Journal of Research on Technology in Education, 37*(3), 259–280.

Leonard, J., Mitchell, M., Barnes-Johnson, J., Unertl, A., Outka-Hill, J., Robinson, R., & Hester-Croff, C. (2018). Preparing teachers to engage rural students in computational thinking through robotics, game design, and culturally responsive teaching. *Journal of Teacher Education, 69*(4), 386–407. https://doi.org/10.1177/0022487117732317

National Research Council. (2011). *Research training in the biomedical, behavioral, and clinical research sciences.* The National Academies Press.

Newton, K., Leonard, J., Buss, A., Wright, C., & Barnes-Johnson, J. (2020). Learning with robotics and game design in an urban context. *Journal of Research in Technology Education, 52*(2), 129–147. https://doi.org/10.1080/15391523.2020.1713263

Papert, S. (1980). *Mindstorms: Children, computers, and powerful ideas.* Basic Books.

Papert, S. (1993). *Mindstorms: Children, computers, and powerful ideas* (2nd ed.). Basic Books.

Pólya, G. (1973). *How to solve it: A new aspect of mathematical method.* Princeton University Press.

Repenning, A., Webb, D., & Ioannidou, A. (2010). Scalable game design and the development of a checklist for getting computational thinking into public schools. In

Proceedings of the 41st AMC technical symposium on computer science education (pp. 265–269). https://doi.org/10.1145/1734263.1734357

Repenning, A., Webb, D. C., Koh, K. H., Nickerson, H., Miller, S. B., Brand, C., Her Many Horses, I., Basawapatna, A., Gluck, F., Grover, R., Gutierrez, K., & Repenning, N. (2015). Scalable game design: A strategy to bring systematic computer science education to schools through game design and simulation creation. *ACM Transactions of Computer Education, 15*(2), 11.1–11.31.

Rich, P. J., & Langston, M. B. (2016). Computational thinking: Toward a unifying definition. In J. M. Spector, D. Ifenthaler, D. G. Sampson, & P. Isaias (Eds.), *Competencies in teaching, learning, and educational leadership in the digital age: Papers from CELDA 2014* (pp. 229–242). Springer.

Ryoo, J. J. (2019). Pedagogy that supports computer science for all. *ACM Transactions on Computing Education, 19*(4), 36:1–36:23.

Schunk, D. H. (2020). *Learning theories: An educational perspective* (8th ed.). Pearson.

Scott, K. A., Sheridan, K. M., & Clark, K. (2015). Culturally responsive computing: A theory revisited. *Learning, Media, & Technology, 40*(4), 412–436. https://doi.org/10.108 0/17439884.2014.924966

Sengupta, P., Kinnebrew, J. S., Basu, S., Biswas, G., & Clark, D. (2013). Integrating computational thinking with K-12 science education using agent-based computation: A theoretical framework. *Educational Information Technologies, 18*(2), 351–380. https://doi.org/10.1007/s10639-012-9240-x

Sullivan, F. R., & Heffernan, J. (2016). Robotics construction kits as computational manipulatives for learning in the STEM disciplines. *Journal of Research on Technology in Education, 48*(2), 105–128. https://doi.org/10.1080/15391523.2016.1146563

Syslo, M. M. (2015, June). From algorithmic to computational thinking [Keynote]. In *Proceedings for the 2015 AMC Conference on Innovation and Technology in Computer Science Education* (p. 1). https://doi.org/10.1145/2729094.2742582

Thomas, J., Joseph, N. M., Williams, A., Crum, C., & Burge, J. (2018). Speaking truth to power: Exploring the intersectional experiences of Black women in computing. In J. Payton, G. K. Thiruvathukal, J. Burge, F. Stukes, Y. Rankin, & E. Dillon (Eds.), *Proceedings of the Research on Equity and Sustained Participation in Engineering, Computing, and Technology* (pp. 36–42). Curran Associates.

Thomas, J. O., Minor, R., & Odemwingie, O. (2016). Exploring African-American middle school girls' perceptions of computational algorithmic thinking of themselves as game designers. In C. K. Loii, J. L. Polman, U. Cress, & P. Reimann (Eds.), *Transforming learning, empowering learners: The International Conference of the Learning Sciences Proceedings, Vol. 1* (pp. 960–962). International Society of the Learning Sciences.

Thomas, J. O., Odemwingie, O. C., Saunders, Q., & Watlerd, M. (2015). Understanding the difficulties African-American middle school girls face while enacting computational algorithmic thinking in the context of game design. *Journal of Computer Science and Information Technology, 3*(1), 15–33.

WBTV Web Staff. (2020, March 22). Charlotte Latin students raise more than $52K to print 3D face shields for medical personnel. WBTV. https://www.wbtv.com/2020/03/22/charlotte-latin-students-raising-money-print-d-face-shields-local-medical-personnel-treat-covid-patients/

Wigfield, A., & Eccles, J. S. (2000). Expectancy-value theory of achievement motivation. *Contemporary Educational Psychology, 25*(1), 68–81. https://doi.org/10.1006/ceps.1999.1015

Wing, J. M. (2006). Computational thinking. *Communications of the ACM, 49*(3), 33–35. https://doi.org/10.1145/1118178.1118215

Wing, J. (2008). Computational thinking and thinking about computing. *Philosophical Transactions of the Royal Society A: Mathematical, Physical and Engineering Sciences, 366*(1881), 3717–3725. https://doi.org/10.1098/rsta.2008.0118

Wing, J. M. (2010). *Computational thinking: What and why?* [Unpublished manuscript]. Computer Science Department, Carnegie Mellon University. https://www.cs.cmu.edu/~CompThink/resources/TheLinkWing.pdf

Yadav, A., Good, J., Voogt, J., & Fisser, P. (2017). Computational thinking as an emerging competence domain. In M. Mulder (Ed.), *Competence-based vocational and professional education* (pp. 1051–1067). Springer.

Yoshida, T. (Series Creator). (1967–1968). *Speed racer.* [TV Series]. Tatsunoko Production; Trans Lux.

2

DESIGNING LEARNING ENVIRONMENTS THAT SUPPORT DEVELOPING COMPUTATIONAL ALGORITHMIC THINKING CAPABILITIES

Jakita O. Thomas

I like the problems [game design for social change] addresses, and I like the fact that it can be applied to daily life. R. King—SCAT Scholar, aged 12

…[game design] allows you to bring new concepts and stories to the table. T. Smith—SCAT Scholar, aged 15

Overall, [SCAT] is one of the greatest experiences I've ever been exposed to as an African-American woman. J. Jones—SCAT Scholar, aged 18

At the beginning of SCAT, I didn't understand why I was there. Then I thought about what I was doing. I was an African-American girl learning how to properly learn game design. As I grew over the years in game designing, I gained a strong liking. The SCAT program has gifted me with a new hobby that most women don't have, and for that I am grateful. L. Washington—SCAT Scholar, aged 18

Computational algorithmic thinking (CAT) is the ability to design, implement, and assess the implementation of algorithms to solve a range of problems. Making a critical aspect of computational thinking explicit, CAT involves identifying and understanding a problem, articulating an algorithm or set of algorithms in the form of a solution to the problem, implementing that solution in such a way that it solves the problem, and evaluating the solution based on some set of criteria (Thomas, 2018; Wing, 2006, 2008, 2010). Furthermore, intersectional computing is defined as "a more complex understanding of the experiences of marginalized groups in computing who live at various intersections of racism, sexism, classism, xenophobia, heterosexism, ableism, etc." (Thomas, 2018, p. 1).

CAT focuses specifically on how the human, as computing agent, engages in the design, implementation, and assessment of an algorithm or set of algorithms to solve a problem on the journey toward choosing the "right" abstractions (Schneider & Gersting, 2019; Thomas, 2015, 2018; Wing, 2008). CAT is an important scaffolded on-ramp as students develop more advanced computational thinking capabilities and apply computational thinking to solve problems that are more constrained and require greater and greater expertise. While having roots in mathematics, through problem solving and algorithmic thinking, CAT has applicability in numerous areas from cooking to music to writing. CAT lies at the heart of computer science, embodying one's ability to think critically and creatively to solve problems (International Society for Technology in Education [ISTE], 2020; Kramer, 2002; Pólya, 1945; Schoenfeld, 2010; Wing, 2006, 2010). Computer science is defined as the "study of algorithms" (Schneider & Gersting, 2019). In classrooms and more "academic" settings, students experience challenges and difficulties enacting CAT (Thomas et al., 2017b), even though, as humans of all ages, we enact CAT in our day-to-day non-academic activities (Bundy, 2007; Wing, 2006; Yadav et al., 2011).

Supporting Computational Algorithmic Thinking (SCAT) is described as a longitudinal between-subjects research project that explores the development over time of CAT capabilities in African-American middle school girls in the context of game design, and a free enrichment program designed to expose African-American middle school girls to game design. SCAT's epistemological orientations centered the experiences of Black girls and were rooted in Black feminist thought and intersectionality (Collins, 2000; Collins & Bilge, 2016; Crenshaw, 1991). As such, SCAT was a learning environment where design, cognition, and identity intersected to create a "Wakanda Effect" (*Black Panther*, Coogler, 2018) situated in Black excellence (Joseph & Thomas, 2020).

The goals of SCAT are as follows: (a) to explore the development of CAT capabilities over three years in African-American middle school girls as they engage in iterative game design and (b) to increase the awareness of participants to the broad applicability of CAT across a number of industries and career paths. Participants, called SCAT Scholars, develop CAT capabilities as they engage in the game design cycle to design more and more complex games over a time period of three years (sixth grade to eighth grade). Each year (called a SCAT Season), SCAT Scholars engage in the following three activities: (a) a two-week intensive game design summer experience; (b) twelve technical workshops focused on implementing the games designed during the two-week summer experience using visual and programming languages (e.g., SCRATCH, App Inventor, Unity) as well as submitting those games to national game design competitions (e.g., National STEM Video Game Challenge, Verizon Innovation App Challenge);

and (c) field trips where Scholars can explore how CAT is enacted in different computing-focused career paths.

This chapter begins to address the research question: How do the individual and small-group CAT capabilities of African-American middle school girls develop over time? We explore and address this question by presenting our findings around the articulation of CAT using data that was collected during SCAT's pilot study. There is little research that focuses on understanding and describing how CAT capabilities are developed over time for any group, but especially for African-American girls and women, whose current representation within the field is incredibly low (Zweben & Bizot, 2017). Therefore, this research fills a gap in the extant literature. We view our articulation of the development of CAT capabilities as a critical first step in describing and identifying changes in the development of CAT capabilities over time for this population. To do so, we created the CAT Capability Flow, which is not only a visual representation of the process(es) scholars engaged in as they used CAT, but also an iteratively refined articulation of the steps involved in CAT.

Perspectives and Theoretical Framework

The National Research Council (NRC, 2011), in their report entitled *A Framework for K12 Science Education: Practices, Crosscutting Concepts, and Core Ideas*, indicates that K12 science and engineering education should be designed so that students "continually build on and revise their knowledge and abilities over multiple years, and support the integration of such knowledge and abilities with the practices needed to engage in scientific inquiry and engineering design" (p. 2). The NRC outlines eight practices as being "essential elements of the K12 science and engineering curriculum" (p. 49). They are: defining problems; developing and using models (physical or mathematical models and prototypes); planning and carrying out investigations; analyzing and interpreting data; using mathematics, information and computer technology, computational thinking; designing solutions; engaging in arguments from evidence; and obtaining, evaluating, and communicating information. Furthermore, the report states that "middle school students should also have opportunities to plan and carry out full engineering design projects in which they define problems in terms of criteria and constraints, research the problem to deepen their relevant knowledge, generate and test possible solutions, and refine their solutions through redesign" (pp. 70–71). The NRC report describes the major competencies that students should have by the 12th grade, and the report also includes sketches describing how those competencies should progress from K12. However, the NRC (2011) notes that those sketches are based on The Committee on a Conceptual Framework for New Science Education Standards' judgment, as "there is very little research evidence as yet on the developmental trajectory of each of these practices" (pp. 3–6).

Research has shown that "since the mid-20th century, computational theories, information and computer technologies, and algorithms have revolutionized virtually all scientific and engineering fields" (NRC, 2011, p. 64). While the definition of computational thinking is still in great debate, Jeanette Wing (2006) offers this definition: "Computational thinking is a way humans solve problems … [it's] solving problems, designing systems, and understanding human behavior, by drawing on the concepts fundamental to computer science" (p. 35). Computational thinking, a fundamental skill involved in analytical thinking, is used by everyone, and is essential to every discipline (Bundy, 2007; Wing, 2006; Yadav et al., 2011). As humans, we engage in computational thinking in many facets of our everyday lives: looking up a name in an alphabetically sorted list, cooking a meal, learning long division, using an automated teller machine (ATM), and standing in line at the supermarket (Wing, 2010). In these scenarios, we as humans develop and/or perform a step-by-step method for accomplishing or analyzing the performance of some task or we design, identify, and/or perform algorithms and assess that design and performance (Schneider & Gersting, 2019). As such, the development and performance of algorithms are critical components of computational thinking. As described previously, we term the ability to design, implement (i.e., perform), and assess the design and implementation of algorithms to solve a range of problems as CAT.

The acquisition and development of skills and practices involve the changing of declarative knowledge to procedural knowledge (Anderson, 2000; Anderson et al., 1981). Declarative knowledge is independent pieces of factual knowledge (Anderson et al., 1981). Procedural knowledge is connected knowledge that forms a process for carrying out a skill (Anderson et al., 1981). A process, if applied in context or among a community, can evolve into a practice (Lave & Wenger, 1991; Nersessian, 2008). While skills or abilities refer to what one can do in the present, capabilities refer to what one can learn to do with instruction and scaffolding (Bandura, 1994; Bransford et al., 1999; Vygotsky, 1978). However, the literature suggests that in order to move learners from capability to ability, learners need four things to be in place: (a) opportunities to make connections between their experiences and the knowledge or skills they are learning; (b) time to develop skills and capabilities so that they can be used at the right time and in the right situations; (c) support as they try to engage in higher level abstraction; and (d) encouragement to engage in metacognitive strategies (Bransford et al., 1999; Owensby, 2006; Thomas, 2008).

As a domain, engaging in game design aligns with the eight practices outlined by the NRC (2011). The game design cycle is an iterative cycle that involves seven phases (Figure 2.1). Each of those seven phases are, themselves, iterative (Fullerton et al., 2004). The brainstorming phase involves designers generating a lot of game ideas, describing the formal and dramatic elements for each idea, and then narrowing those ideas down to a small subset of ideas, one of which

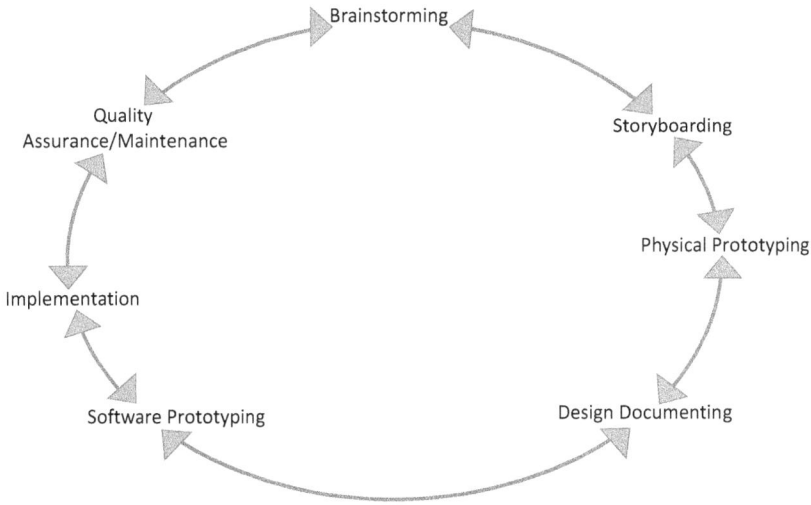

FIGURE 2.1   Game Design Cycle

they choose to pursue. The storyboarding phase involves the creation of a storyboard, which serves as a gameplay treatment, often depicting demo artwork. Next, designers develop a physical prototype, often using craft materials to construct it; then the physical prototype is play-tested by actual players from the target user group who play the game and provide feedback about their gameplay experience as the designer observes in real time (DiSalvo et al., 2009; Fullerton, et al., 2004). Drafting the design document comes next. The design document is a dynamic document that describes every aspect of the game from the way that it will function down to the file types that will be used for game assets such as images, sprites, or music.

Designers then develop rough software prototypes, which are often rapidly constructed to help them, through play-testing, answer questions that may remain about their designs. For example, designers may have a question about whether resources in the game should be made available to players all at once or introduced or "unlocked" as gameplay progresses. In this instance, designers would build out two rough software prototypes, implementing just enough of the game so that both versions can be play-tested and players can provide feedback about which version of the game provides the best player experience (i.e., the version with all of the resources available at once or the version with resources unlocked as gameplay progresses).

The software prototyping phase is then followed by the implementation phase, where designers implement their games using visual or programming languages (i.e., SCRATCH or XNA), and/or a game engine like Unity or Unreal. Finally, quality assurance testing is done to ensure that all of the known bugs in the game are fixed prior to release of the game to the market with maintenance addressing bugs that arise once the game has been released. Quality assurance/maintenance occurs with continued play-testing.

Every game has formal and dramatic elements. Formal elements are defined as elements that all games share (Fullerton et al., 2004). There are eight formal elements: players, objectives, procedures, rules, resources, conflict, boundaries, and outcome. Players are a "voluntary participant or set of participants who both partake in as well as consume the game" (Fullerton et al., 2004). Objectives are the goals the player is trying to reach. Procedures are the actions or methods of play allowed by the rules. The procedures describe who does what, where, when, and how. The rules of a game describe the limits or parameters for what a player can and cannot do. Resources are objects in a game that help players achieve their goals, and they are often scarce in the game system. Conflict involves the actions that keep the player from directly achieving the goals and/or actions that can make players work against each other to accomplish the goals. The boundaries of a game are the ways in which the rules and goals that are driving the game apply within the game compared to "real life." So, for example, if you die within the game context, you don't actually die in real life. Finally, the outcome of a game describes the certainty of a measurable and unequal result that deems some player(s) winner(s) and others as loser(s).

Dramatic elements engage players by creating a dramatic context for the formal elements (Fullerton et al., 2004). Unlike the formal elements, not all dramatic elements are required in order for a game to be a game. However, including the right dramatic elements in the right way for a particular game generally increases the likelihood that players will find the gameplay experience engaging and fun. There are five basic dramatic elements: challenge, play, premise, character, and story. Challenge describes the tensions created as the player(s) work to resolve problems or conflict in the game and the varying levels of achievement or frustration they experience as a result. Play involves the "free movement of the player within a more rigid structure" (Fullerton et al., 2004). Turn taking is an example of a particular style of play, as players are typically allowed to do a wide range of things when it's their turn to play and often aren't able to do anything when it isn't their turn. The premise is the overarching context of the game and often the thing that creates the conflict and sets the gameplay in motion. Characters are the agents through which dramatic stories are told. Characters can also be vessels for player participation, allowing the player to "walk in the shoes" of a particular character. Finally, the story is the narrative that unfolds when the game is played. Together, the formal and dramatic elements create a game system, and it is the game designer's task to design that system in such a way that the result is an experience that the player assesses as being fun.

Game design has been leveraged as an approach to broadening participation in computer science, especially among groups with low representation in the field (e.g., women, minorities). The literature describes research that leverages game design to increase student interest and retention in computer science as well as to improve computing capabilities, programming skills, technology fluency, algorithmic thinking, abstraction, and modeling capabilities from K-16 (Barnes et al., 2007; Barron et al., 1998; DiSalvo et al., 2009; Maloney et al., 2004; Repenning & Ioannidou, 2008; Werner et al., 2005; Werner et al., 2009). In other words, these research efforts have impacted students' computational thinking capabilities. Game design has roots in constructionism and project-based learning, affording young people the opportunity to construct/design their own games and "create new relationships with knowledge in the process" (Kafai, 2006, p. 38) both in formal and informal learning environments (Koschmann, 1996; Papert, 1993; Repenning et al., 2010). Our work aims to go one level deeper, exploring how African-American middle school girls develop CAT capabilities over time.

Supporting Computational Algorithmic Thinking

SCAT Scholars began the program the summer prior to their sixth-grade year and continued through their eighth-grade year (i.e., June 2013–May 2016). Each year, SCAT Scholars engaged in the following three activities: (a) a two-week intensive game design summer experience; (b) twelve technical workshops focused on implementing the games designed during the two-week summer experience using visual and programming languages and industry-standard game engines (e.g., SCRATCH, App Inventor, Unity) as well as submitting those games to national game design competitions (e.g., National STEM Video Game Challenge, Verizon Innovation App Challenge); and (c) field trips where Scholars could explore how CAT is enacted in different computing-focused career paths. Scholars were supported by several scaffolds in the learning environment, including the facilitator, four undergraduate assistants, the Design Notebook, and other Scholars, all providing support in the ways that cognitive apprenticeship suggests (Collins et al., 1989; Thomas, 2014; Thomas et al., 2016; Thomas et al., 2017).

Facilitator as Modeler and Coach

In the SCAT learning environment, the facilitator played a major role in supporting Scholars as they developed their CAT capabilities over time. She served primarily as a role model within the learning environment, leading and supporting the whole group and dyad discussions as they thought through and iteratively refined their designs. Specifically, she helped Scholars make connections across dyads, modeling the kinds of questions they should be asking themselves and each other along the way (Lave & Wenger, 1991; Thomas, 2015). The facilitator served, secondarily, as a just-in-time coach, walking from group to group asking dyads questions about their design decisions, uncovering issues with their design, showing them how to use the Design Notebook and other resources within the SCAT environment as they designed their games, and providing a listening ear as they designed—in essence, serving as a sounding board (Thomas, 2015).

The Design Notebook in Support of CAT

While the facilitator was a critical component of the SCAT learning environment, it was not physically possible for her, or the undergraduate assistants, to be with every group or individual at all times. To help overcome that limitation, the Design Notebook was developed and integrated into SCAT activities to coach Scholars' enactment of CAT as they moved through the game design cycle; it allowed Scholars to develop more expert CAT capabilities over time in the ways that cognitive apprenticeship suggests using a system of scaffolds (Collins et al., 1989; Owensby, 2006; Roschelle, 1996; Thomas et al., 2015; Vygotsky, 1978). Each scaffold in the system, which has five parts, supported groups and individuals in a particular way, addressing difficulties that learners encountered when enacting complex cognitive skills, processes, and capabilities, that is, designing experiments, engaging in case use, or enacting CAT (Owensby, 2006; Thomas, 2008; Thomas et al., 2017). First, tools are structured in such a way that they suggest the process that learners are engaging in at a high level by making the process sequence visible. Second, within each tool as learners are carrying out a task or reflecting on a task, structured questioning or statements, called prompts, make the task sequence clear and focus learners' attention. Third and fourth, for each prompt in the sequence, hints and examples are provided. Hints are task- or domain-specific statements or questions used to refine a task, while examples model a process or a particular step of a process. Fifth, a template or chart is provided for some tasks in the sequencing to help learners line up their reasoning.

Other Scholars

Scholars could move through the iterative game design cycle at their own pace. As a result, it was likely that those Scholars or dyads who were further along in the game design cycle would be able to scaffold the work of the dyads who were not as far along (Owensby, 2006; Repenning & Ioannidou, 2008; Schoenfeld, 2010; Thomas, 2008; Wells & Chang-Wells, 1992). Additionally, the different perspectives that Scholars brought into the SCAT learning environment contributed to greater understanding by the dyad through small group collaboration and discussion, the benefits of which are well documented in the literature (Barnes et al., 2007; Barron et al., 1998; Bransford et al., 1999; Lave & Wenger, 1991; Nersessian, 2008).

Computational Algorithmic Thinking in SCAT

As described earlier the game design cycle itself is an algorithm of seven phases, and Scholars engaged in CAT as they moved through and between the phases. Throughout the game design cycle, CAT was enacted at numerous points

(Thomas et al., 2017). For example, during the storyboarding phase, Scholars engaged in CAT as they drew stills that depicted both the visual and non-visual elements of gameplay as the player moves through the game from beginning to end. In fact, the storyboard was the first enactment of CAT, visually describing many of the game's algorithms. During the physical prototyping phase, Scholars articulated the primary algorithms that govern gameplay as well as how the players engage with the game. They did this by articulating the rules and procedures of the game. The implementation phase involved Scholars articulating algorithms in SCRATCH to implement gameplay functionality and behavior as well as adapting and implementing common SCRATCH algorithms (e.g., creating a scrolling screen, keeping score, enacting a timer, detecting and responding to a collision between objects). During each phase, Scholars were tasked with articulating the algorithms in their games in increasingly specific ways with the result being a fully functional game.

Methods

The CAT Capability Flow should be thought of as a first step in describing a process or capability that has not been described in detail before for any population, but especially for African-American middle school girls. It should not be thought of as an absolute or complete description of CAT. Leveraging Owensby (2006), we anticipate that the CAT Capability Flow will, ultimately, inform potential developmental changes that may occur in Scholars' ability to design, implement, and assess algorithms over time, which is outside the scope of this chapter. This chapter does describe the first step of this process: constructing the CAT Capability Flow (Owensby, 2006).

The Pilot Study

We piloted the SCAT Workshops and Game Design Competition activity from January 2013 to March 2013. One major purpose of this eight-week pilot study was to begin identifying and describing the sub-processes of CAT. During the pilot study, Scholars iteratively designed and implemented a game for social change, which was play-tested along the way. It is important to note that data collection for the full implementation of the SCAT project took place from July 2013 to May 2016, with full data analysis beginning in June 2016 and continuing to date. This is due to the volume of data we have collected as well as our commitment to analyzing that data carefully, especially video observations. (There are more than 2,600 hours of video observation data.) Therefore, analysis has taken a great deal of time. We started from the beginning, or the pilot study, to ensure that we describe CAT capability development over time in the most complete and appropriate way.

Setting and Participants

Our pilot study took place at a private, liberal arts, Historically Black College/ University (HBCU) located in the Southeastern United States. Over the course of the pilot study, we engaged 20 African-American sixth-grade girls (SCAT Scholars) from across the metro-Atlanta area in the design and implementation of a game for social change. Each Saturday for eight weeks, Scholars met for 6 or 7 hours to design games for social change (two Saturdays) and to implement those games (six Saturdays).

Data Sources

A number of different types of data were collected over the course of the SCAT project including: whole class as well as small group video observations, written direct observations, pictures, Design Notebook pages, and dyad artifacts such as physical prototypes, design documents, and implementation files of dyads' games. To address the research question posed at the beginning of this chapter and to inform the creation of the CAT Capability Flow, we focused particularly on direct observations, whole class video observations, and pictures from the pilot study. These data helped us understand how the game design cycle was enacted.

Data Analysis

Six members of the research team used the data described earlier during data analysis. Data analysis involved watching the whole class video observations (more than 64 hours of video data) of the eight-week enactment (Thomas, 2018). Using the video observations and pictures, for every phase in the game design cycle, except for the software prototyping and quality assurance/maintenance phases, two members of the research team individually documented and described that phase's enactment. The descriptions that resulted were discussed by the research team. Leveraging the resulting descriptions, the research team then mapped our high-level definition of CAT, described earlier, to each phase of the game design cycle. Following that mapping, the activities conducted during each phase were examined to identify the order in which activities occurred within a phase, either serially or in parallel. This was accomplished by team members independently watching the video observations as well as reviewing the observation notes that the research team collected during that phase's enactment and then convening to discuss. It is important to note that some steps within a phase could have occurred in either random order or in parallel. For example, as described in Thomas (2018), during the brainstorming phase, Scholars generated ideas and/or adapted known games (steps 1 and 2); narrowed their ideas down to at least three (step 3); described the characters, formal elements, and dramatic elements in their ideas (steps 4, 5, and 6), and created PowerPoint presentations for their top three ideas (step 7), which they presented to the larger group (step 8). We repeated this process for each of the remaining three phases of the game design cycle enacted during the pilot study (i.e., physical prototyping, design document, implementation) as well as play-testing, which is not a formal phase but an integral part of game design described earlier.

Limitations of the Study

The first limitation involves the small number ($n = 20$) of participants in the study (i.e., African-American middle school girls). We kept the participant size small for several reasons. First, we wanted to understand whether the number of undergraduate assistants was sufficient for a group of this size. We engaged four undergraduate assistants to work with the group, which meant that any one undergraduate assistant would have at most six or seven groups to provide just-in-time coaching as needed. Undergraduate assistants also rotated observation duties, which meant that the remaining undergraduate assistants had to distribute the groups for which they would provide just-in-time coaching. Second, we anticipated that the amount of data gathered, especially the video observations, would be large. As a result, to make the analysis of the video data somewhat manageable, we had to keep the participant size low. Additionally, because this work is so exploratory, we felt that the small number of participants was reasonable. Finally, we are not able to make generalizable claims about whether the CAT Capability Flow that resulted in this population (i.e., African-American middle school girls) will be the same for other groups (i.e., African-American middle school boys, girls or boys of other races and/or ethnicities). However, again, given that this work is exploratory, this chapter represents an initial stake in the ground for a group that has incredibly low representation in computer science and in the game design industry—African-American girls and women.

Results

Figure 2.2 shows the mapping of the game design cycle onto our high-level definition of CAT described earlier. For example, each phase of the game design cycle required Scholars to describe the game in more specific ways (Thomas et al., 2017). As a result, every phase involved some aspect of "understand the problem" (Thomas et al., 2017). During the brainstorming phase, Scholars described the formal and dramatic elements of their game, providing a very high-level description of the game itself. During the storyboarding phase, Scholars described the formal and dramatic elements visually, depicted how the characters and objects within the game would look, how they would move, what the game world looked like, what resources would be available in the game and where those resources would be in the game world, the perspective of the player (e.g.,

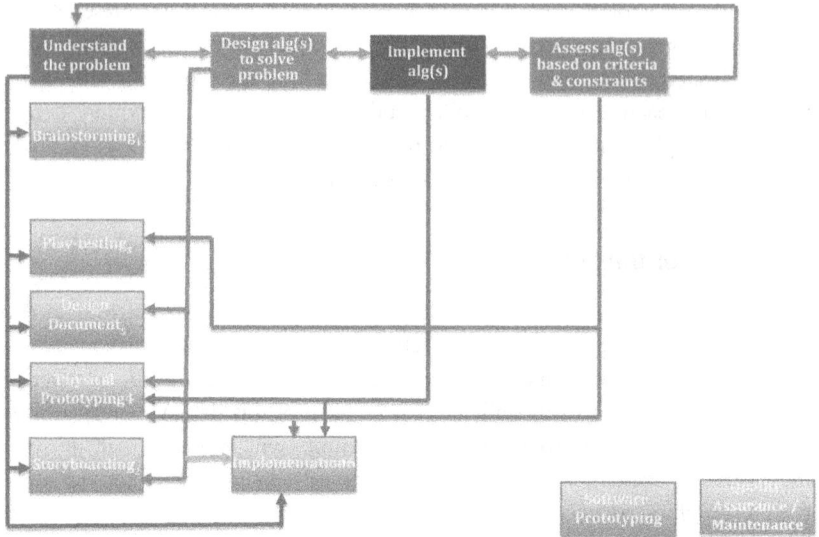

FIGURE 2.2 CAT Capability Flow Version 1: Mapping CAT Definition and Game Design Cycle

first person, third person), how the player would move, and any other non-visual elements that would enhance the player experience (e.g., music). During the physical prototyping phase, Scholars used craft materials to represent gameplay and the core gameplay mechanic in the physical world.

During this phase, Scholars had to further define and describe how characters, resources, objects and the game world looked as well as how the actions of the player informed the interactions between players, players and objects, as well as between objects within the game. They also had to clearly define and measure player progression through the game toward the goal. Additionally, Scholars had to describe how the game would reward players as they made progress toward the goal or objective, as well as how the game would penalize players as they made progress away from the goal or objective.

As Scholars implemented their games in SCRATCH, they had to translate their designs to represent the game world, sprites, and look of their game in SCRATCH as well as the functionality of their games by creating blocks that described how player actions changed the behavior of sprites within the game world. During play-testing sessions, Scholars played each other's games and gave feedback. As Scholars reviewed and discussed feedback they received during play-testing sessions, they were able to better understand how to convey the social issue they addressed in their games to players in such a way that players would be engaged and motivated to play the game to achieve the game's objective(s) and, ultimately, become more aware about the social issues that Scholars' games addressed.

Figure 2.2 allowed us to understand which phases of the game design cycle addressed various aspects of CAT. For example, during the physical prototyping phase, dyads understood the problem they were trying to address through their games better because building the physical prototype forced them to be more specific about their idea as they represented each aspect of the game using physical media. Dyads also designed algorithms to solve the problem in the form of rules and procedures, implemented those algorithms by creating a physical representation of the game world and providing written instructions for how to play the game, and assessed those algorithms by having their peers play-test their games and give feedback on the physical representation as well as the gameplay. In addition, dyads also collaboratively assessed the feedback given by their peers during play-testing sessions to inform design decisions for and changes to their games. The software prototyping and quality assurance/maintenance phases are not connected to the rest of the flow because these phases were not included in the pilot study enactment due to time constraints. It should be noted that, in the figures, each phase has a subscript: (a) brainstorming, (b) play-testing, (c) storyboarding, (d) physical prototyping, (e) design document, and (f) implementation. These subscripts will be used later as the enactment for each phase is described.

Specifying Each Phase of the Game Design Cycle

The prioritized set of activities that Scholars engaged in during both the brainstorming phase and the storyboarding phase are described in full detail in Thomas (2018). In this chapter, we include Figures 2.3 and 2.4, which depict the prioritized enactment of the brainstorming and storyboarding phases, respectively, substituted for the brainstorming and storyboarding blocks depicted in Figure 2.2. We include these figures in this chapter for completeness, but we will begin our description of the prioritized enactments in this chapter for the remaining three phases—physical prototyping, design document, and implementation—that were enacted during the pilot study as well as play-testing.

Play-Testing/Gallery Walk

As described previously, play-testing involves bringing actual players from the target user group in and observing them as they play the game in real time, getting feedback about the game experience to iteratively inform the design of the game. For SCAT, we employed a pared down play-testing enactment called a "gallery walk" (Kolodner, 2002). A gallery walk is an informal presentation of an iteration of a student or Scholar-created artifact in which peers review the artifact, providing feedback that will inform the next iteration. The gallery walk affords "time for engaging publicly in debugging, explaining, and redesigning" (Kolodner, 2002). Gallery walks are followed by whole group discussions facilitated by the teacher or facilitator to help students or Scholars review the

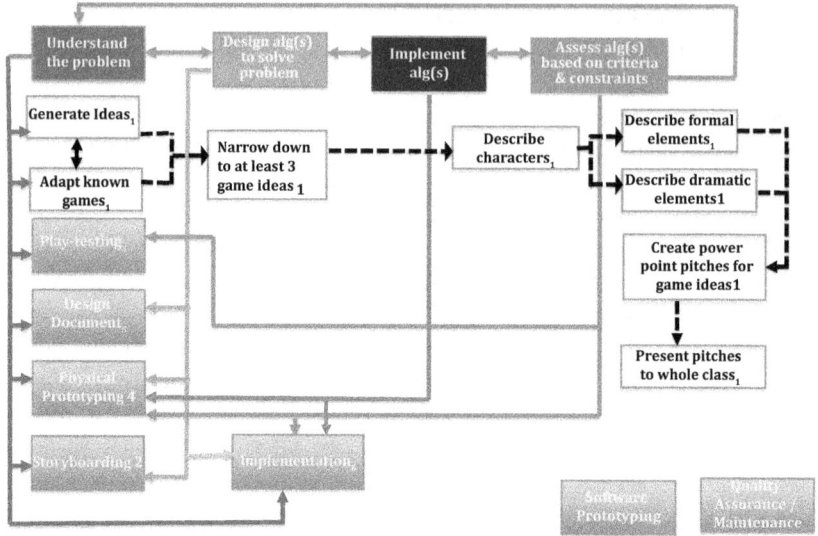

FIGURE 2.3 CAT Capability Flow with Fleshed Out Brainstorming Phase

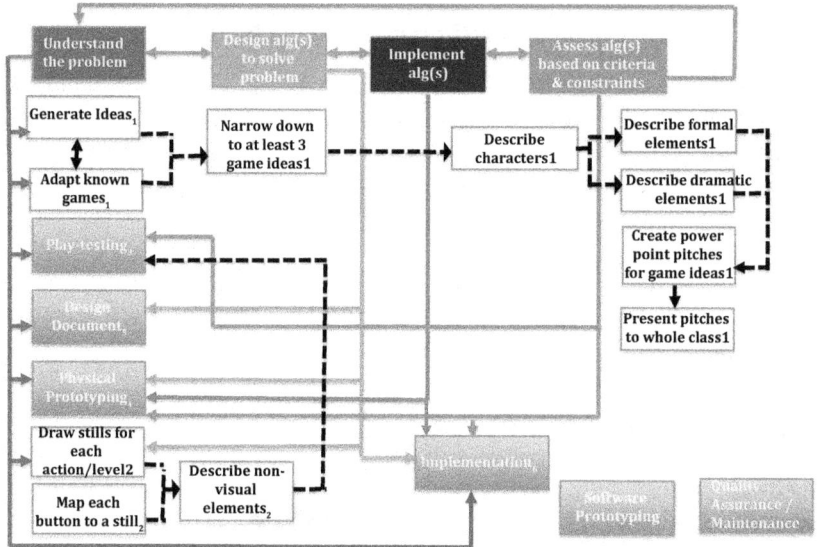

FIGURE 2.4 CAT Capability Flow with Fleshed Out Brainstorming and Storyboarding Phases

feedback they received, make sense of it, ask questions about it, and make decisions about changes to make and things to add or remove from the artifact's design. For SCAT, dyads used the Play-Testing/Gallery Walk Design Notebook Page to gather feedback from their peers (Figure 2.5).

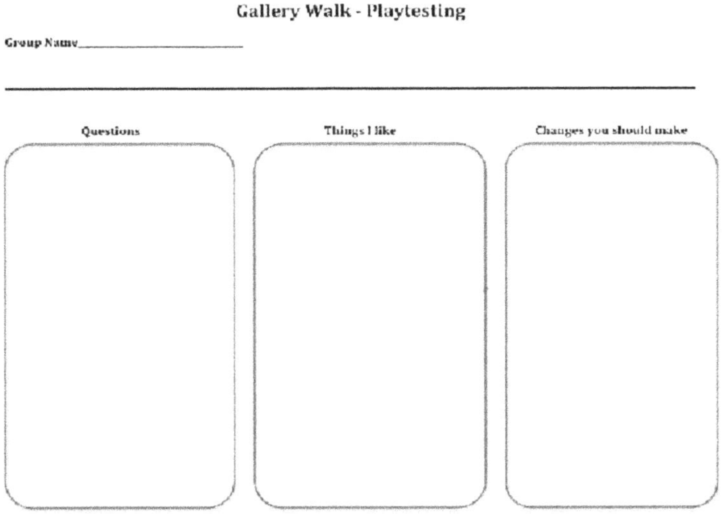

Gallery Walk - Playtesting

Group Name_____

Questions	Things I like	Changes you should make

FIGURE 2.5 Play-Testing/Gallery Walk Design Notebook Page

The Play-Testing/Gallery Walk Design Notebook Page prompted Scholars to give particular kinds of feedback about the games they gallery walked or play-tested, namely: questions that arose as they played, which often represented aspects of the game design that were unclear; things they liked about the game; and changes they would suggest the game designers make in order to address aspects of the game design that were unclear or that made the game less engaging or fun. During a gallery walk, Scholars played the games in step 1. Then, they either engaged in steps 2–5 (see Figure 2.5) to describe aspects of the game that were not understood, described what they liked about the game, described what they did not like about the game, and/or described potential changes the game designers should make. As game designers, during play-testing/gallery walk, Scholars either engaged in steps 1–3 (see Figure 2.5), played their game (i.e., self-testing), observed what play-testers did while playing the game, and/or gathered feedback. However, these three activities could happen concurrently. After play-testing/gallery walk feedback were gathered, dyads engaged in step 4 (see Figure 2.5) and reflected on that feedback. Figure 2.6 shows this prioritized enactment of the play-testing phase substituted for the play-testing block depicted in Figure 2.2.

Physical Prototyping Phase

During the physical prototyping phase, dyads built physical models of their games using craft materials. Figure 2.7 shows an example of a physical prototype. The purpose of a physical prototype was two-fold. First, the physical prototype

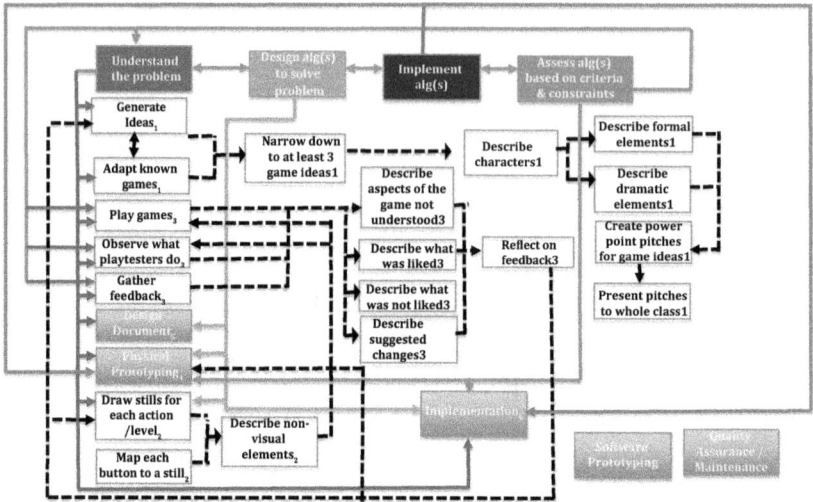

FIGURE 2.6 CAT Capability Flow with Fleshed Out Brainstorming, Storyboarding, and Play-Testing Phases

helped game designers articulate the details of the design of the game in an even more specific way than the two phases which preceded it (i.e., brainstorming and storyboarding). Second, the physical prototype helped Scholars methodically explore the efficacy of the core game mechanic, which is typically defined by the action that players perform most repeatedly during gameplay in order to achieve the goal or objective of the game (Fullerton et al., 2004).

During the physical prototyping phase, dyads created initial visualizations of the game (step 1), which included describing the worlds or contexts in which gameplay would take place inside of the games, the actions the player would take during gameplay, and the ways in which the game would measure and/or reward the impact of those actions on the player's progress toward achieving the game's objective. Next, dyads built the structure and foundation of the game (step 2), which involved creating the physical model of the game. Finally, dyads either completed steps 3–5 (see Figure 2.7), methodically identified and separated the rules of the game from the features of the game ('*Separate rules from features*'), created or refined the full set of rules, and/or created or refined the procedures of the game. Figure 2.8 shows this prioritized enactment of the physical prototyping phase substituted for the physical prototyping block depicted in Figure 2.2.

Design Document Phase

In the design document phase, dyads created a written description of their game, describing the gameplay, look, and functionality of their games at an even more fine-grained level than was required in the previous phases. Because dyads were

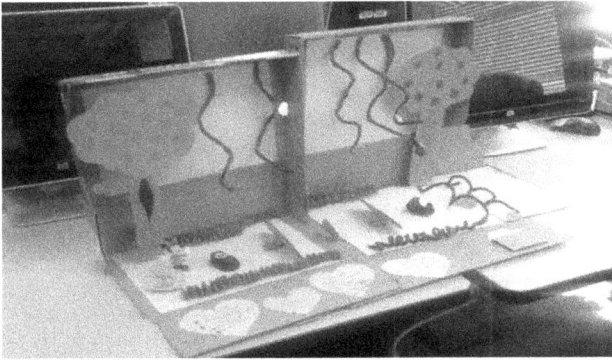

FIGURE 2.7 SCAT Dyad's Physical Prototype

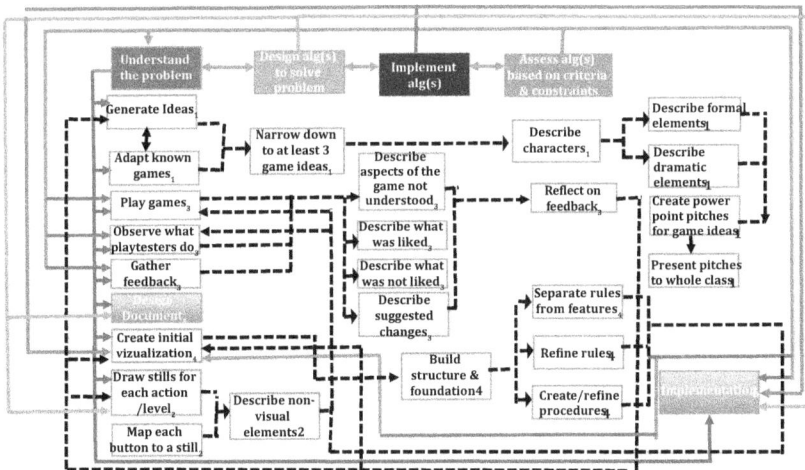

FIGURE 2.8 CAT Capability Flow with Fleshed Out Brainstorming, Storyboarding, Play-Testing, and Physical Prototyping Phases

working to submit either their implemented game or their design document to the National STEM videogame challenge, we used the template provided by the challenge organizers on their website as the model for the dyads' design documents. The current template that is on their website is different from the one we used that was on their website in 2013. Table 2.1 describes the prompts in the 2013 template. The facilitator and undergraduate assistants scaffolded dyads as they worked section by section using Microsoft Word to complete their design documents.

During this phase, dyads created their title page including the name of their game and members of their team (step 1). Next, dyads developed their team bio, featuring pictures of themselves, their names, the schools they attended, and

TABLE 2.1 Prompts Scholars Used to Develop Their Design Documents

Topic	Description
Overall Vision for the Game	This section should provide a short summary or description of the game.
Target Audience	Great game designers always design their games with a specific audience in mind, and this section should describe that audience.
Platform	Is the game designed to be played on a game console? A mobile device? The web? A good game design targets a specific platform and uses the capabilities of that platform to its advantage.
Genre	This section should describe the genre of the game. Popular genres include action, adventure, sports, strategy, puzzle, racing, platformer, and role-playing.
Core Gameplay	This section should describe in detail what playing the game is like including core game mechanics, goals, components, controls, and user experience.
Visual Style	This section should describe the look and feel of the game.
Characters and Storyline (if applicable)	This section should provide a short summary or description of the game. Imagine you are pitching the game to a friend while riding on an elevator. How would you describe the game in a minute or less?

the program that they were participating in (step 2) (i.e., SCAT). Then, dyads described the background story of their game with particular focus on the issue or problem they wanted their game to address or bring awareness to (step 3). Dyads identified the platform that their game would run on (step 4) and identified the genre of game they were designing (step 5). Next, dyads described the goals of the game (*'State game's goals'*) (step 6). Then, dyads described key points of the game (step 7), described player interactions, including the core gameplay and visuals of the game (step 8), and described levels of gameplay for their games, providing an overview of what the player would be aiming to achieve on each level of the game (step 9). Finally, dyads described how the player would advance from level to level (step 10). Figure 2.9 shows this prioritized enactment of the design document phase substituted for the design document block depicted in Figure 2.2.

Implementation Phase

During the implementation phase, dyads created a functional software version of their games. They used SCRATCH as the programming environment for implementation. SCRATCH is a useful programming environment for understanding CAT because the focus of SCRATCH is not on the syntax, or rules, of a programming language (although syntax is conveyed, as only certain types of blocks will fit together), but more about the semantics, or meaning, that its users

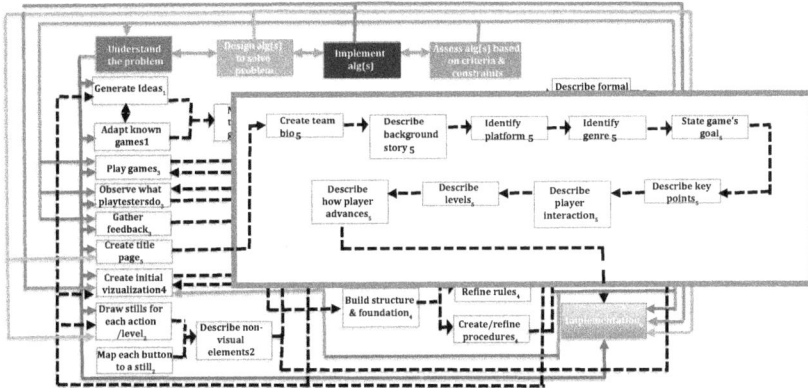

FIGURE 2.9 CAT Capability Flow with Fleshed Out Brainstorming, Storyboarding, Play-Testing, Physical Prototyping, and Design Document Phases

are trying to convey as they put blocks together to create algorithms in their games. Accordingly, the sets of blocks that dyads put together in SCRATCH to represent the behavior and functionality of their games represent two different kinds of algorithms. First, the sets of blocks represent the algorithms of the games as dyads essentially use SCRATCH to implement the rules and procedures of their games (Thomas et al., 2017). Second, the sets of blocks represent dyads' understanding of common SCRATCH algorithms as many of those common SCRATCH algorithms were present in their games (i.e., implementing a timer, changing levels, making a sprite move) (Thomas et al., 2017).

Figure 2.10 shows an example of a subset of the algorithms defined by one SCAT dyad through the construction of blocks in SCRATCH. For example, the set of blocks in the upper left-hand side of the figure indicate that if the player touches the blue line (designated as Sprite1), then the player's character (i.e., the girl in the white pants) should move back to her starting position in the upper left-hand corner of the game screen shown in the square on the upper right-hand side of Figure 2.11.

Scholars spent six weeks implementing their games, meeting for 6 or 7 hours on Saturday mornings. During the implementation phase, dyads constructed blocks to enact the functionality they designed their games to have ('*Construct blocks*') (step 1). Next, they self-tested their games, running them to identify any errors or changes that they should make to the game prior to play-testing ('*Self-test*') (step 2). After self-testing, dyads debugged their games to correct any errors and made changes and refinements to their games based on the outcomes of the self-testing ('*Debug/Assess algs*') (step 3). After self-testing, dyads were ready for the larger SCAT group to play-test their games. Figure 2.11 shows this prioritized enactment of the implementation phase substituted for the implementation block depicted in Figure 2.2.

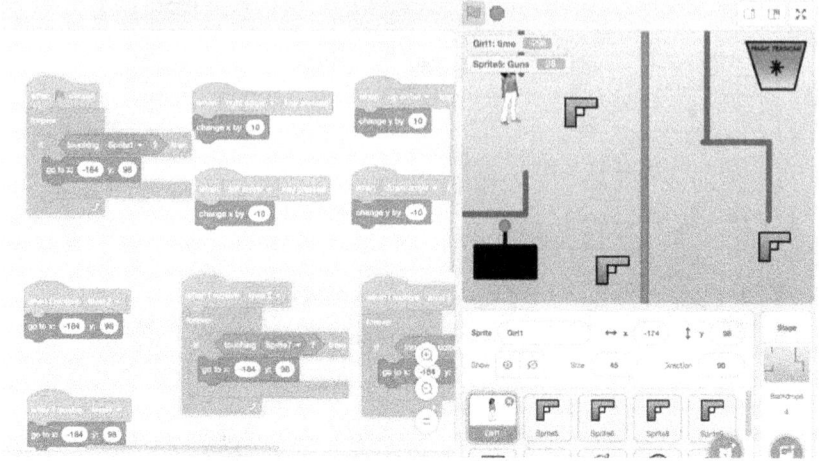

FIGURE 2.10 Screenshot from Dyad's Game Called "Goodbye Guns!"

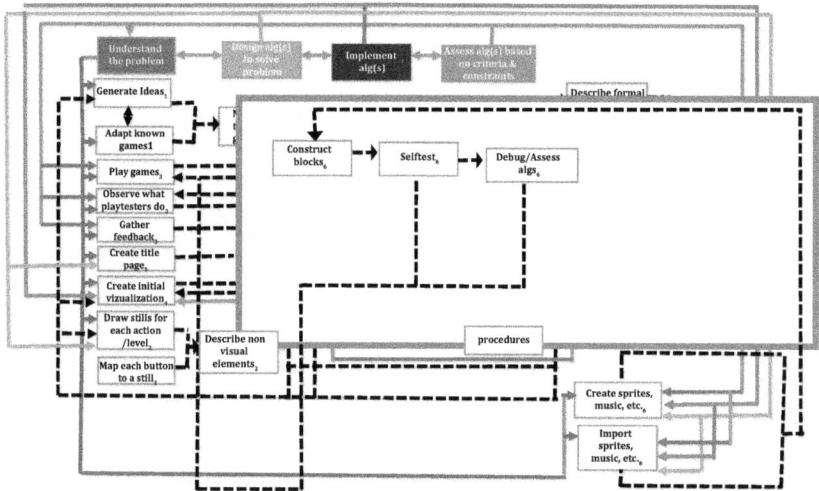

FIGURE 2.11 CAT Capability Flow with Fleshed out Brainstorming, Storyboarding, Play-Testing, Physical Prototyping, and Implementation Phases

The iterative refinement of the CAT Capability Flow will continue as our analysis of video observations, direct observation notes, and picture data collected during SCAT Seasons 1, 2, and 3 continues, resulting in a more complete picture of what is involved in CAT in the context of game design. Note that the video observations include more than 2,600 hours of whole group and dyad video content for SCAT Seasons 1, 2, and 3. That analysis will also include the software prototyping phase of the game design cycle, since that phase was enacted during SCAT Seasons 1, 2, and 3. As none of the dyads has released their games commercially to date, the quality assurance/maintenance phase will not be a part of the future CAT Capability Flow as a result of data collection and analysis for the pilot study and SCAT seasons 1, 2, and 3.

Discussion

Our analysis of the CAT Capability Flow to date reveals some compelling insights. First, enactment of CAT in the context of game design requires significant effort from students. Recall that the game design cycle is an iterative cycle of phases which are themselves iterative. The CAT Capability Flow represents both the serial, parallel, and iterative nature of the enactment of CAT in the context of game design. Second, our analysis reveals that different aspects of CAT show up in different phases of the game design cycle for this population (i.e., African-American middle school girls). For example, while "understand the problem" maps to every phase of the game design cycle, "implement algorithm(s)" (described as *"Implement alg(s)"*) maps to the physical prototyping and implementation phases only. Prior research reveals that subsequent phases of the game design cycle require the game designers (i.e., Scholars working in dyads) to describe their games in increasingly specific ways (Thomas et al., 2017). This increased specification over subsequent phases could result in a deeper understanding not only of the game, but also potentially of the problem or social issue itself. Analysis of the dyad video observations will reveal whether dyads discussed or conducted further research on their social issue as they moved through subsequent phases of the game design cycle.

Finally, our analysis revealed two phases where all aspects of CAT were mapped onto, namely, the physical prototyping and implementation phases. While analysis of dyad video observations and dyad artifacts will uncover fully what this suggests about CAT capability development in African-American middle school girls, previous research suggests that the physical prototyping and implementation phases were two phases where dyads reported experiencing the most difficulties as they were designing their games (Thomas et al., 2017). Therefore, our analysis suggests that phases with multiple arrows coming into them may be phases where additional scaffolds are required to support Scholars as they move through those phases. Our continued analysis of dyad video observations will provide more insight into what kinds of additional scaffolds may be needed.

Summary

This chapter describes preliminary findings related to CAT, a complex cognitive capability, which has previously not been articulated or described. We are engaging in an approach that will result in the continuous iterative refinement of the CAT Capability Flow. We are continuing our analysis of the three seasons of data collected during the full implementation of the research, which will result in even more detailed versions of the CAT Capability Flow over the course of the project in the coming years.

References

Anderson, J. R. (2000). *Cognitive psychology and its implications* (5th ed.). Worth Publishing.

Anderson, J. R., Greeno, J. G., Kline, P. J., & Neves, D. M. (1981). Acquisition of problem-solving skills. In J. R. Anderson (Ed.), *Cognitive skills and their acquisition* (pp. 191–230). Lawrence Erlbaum Associates.

Bandura, A. (1994). Self-efficacy. In R. J. Corsini (Ed.), *Encyclopedia of psychology*, Vol. 3 (2nd ed., pp. 368–369). Wiley.

Barnes, T., Richter, H., Chaffin, A., Godwin, A., & Powell, E. (2007). Game2Learn: Building CS1 learning games for retention. In *Proceedings of the 12th annual SIGCSE conference on Innovation and technology in computer science education* (pp. 121–125). ACM.

Barron, B., Schwartz, D. L., Vye, N. J., Moore, A., Petrosino, A., Zech, L., Bransford, J. D. & The Cognition and Technology Group at Vanderbilt. (1998). Doing with understanding: Lessons from research on problem- and project-based learning. *Journal of the Learning Sciences*, 7(3&4), 271–311.

Bransford, J. D., Brown, A. L., & Cocking, R. R. (1999). *How people learn: Brain, mind, experience, and school*. National Academy Press.

Bundy, A. (2007). Computational thinking is pervasive. *Journal of Scientific and Practical Computing*, 1(2), 67–69.

Collins, A., Brown, J. S., & Newman, S.E. (1989). Cognitive apprenticeship: Teaching the crafts of reading, writing, and mathematics. In L.B. Resnick (Ed.), *Knowing, learning, and instruction: Essays in honor of Robert Glaser* (pp. 453–494). Lawrence Erlbaum Associates.

Collins, P. H. (2000). *Black feminist thought*. Routledge.

Collins, P. H., & Bilge, S. (2016). *Intersectionality*. Polity Press.

Coogler, R. (Director). (2018). *Black Panther* [Film]. Marvel Studios.

Crenshaw, K. (1991). Mapping the margins: Intersectionality, identity politics, and violence against women of color. *Stanford Law Review*, 43(6), 1241–1299.

DiSalvo, B. J., Guzdial, M., McKlin, T., Meadows, C., Perry, K., Steward, C., & Bruckman, A. (2009). *Glitch game testers: African American med breaking open the console*. Proceedings of Digital Games Research Association. https://e-channel.med.utah.edu/wp-content/uploads/2016/07/digra2009_49.pdf

Fullerton, T., Swain, C., & Hoffman, S. (2004). *Game design workshop: Designing, prototyping, & playtesting games*. Focal Press.

International Society for Technology in Education [ISTE]. (2020). *ISTE standards*. https://www.iste.org/standards

Joseph, N. M., & Thomas, J. O. (2020, June 19–23). *Designing STEM learning environments to support middle school Black girls' computational algorithmic thinking: A possibility model for disrupting STEM neoliberal projects* [Paper presentation]. International Conference of the Learning Science, Nashville, TN, United States. (Conference canceled).

Kafai, Y. B. (2006). Playing and making games for learning: Instructionist and constructionist perspectives for game studies. *Games and Culture*, 1(1), 36–40.

Kolodner, J. L. (2002). Facilitating the learning of design practices: Lessons learned from an inquiry into science education. *Journal of Industrial Teacher Education*, 39(3), 9–40.

Koschmann, T. (1996). Paradigm shifts and instructional technology: An introduction. In T. Koschmann (Ed.), *CSCL: Theory and practice of an emerging paradigm* (pp. 1–23). Lawrence Erlbaum Associates.

Kramer, K. D. (2002). Algorithms should mind your business. http://www.outsourcing-russia.com/docs/?doc=680.

Lave, J., & Wenger, E. (1991). *Situated learning: Legitimate peripheral participation.* Cambridge University Press.

Maloney, J., Burd, L., Kafai, Y., Rusk, N., Silverman, B., & Resnick, M. (2004, January). Scratch: A sneak preview. In *Proceedings of the Second International Conference on Creating, Connecting and Collaborating through Computing* (pp. 104–109). IEEE.

National Research Council. (2011). *A framework for K–12 science education: Practices, crosscutting concepts, and core ideas.* The National Academies Press.

Nersessian, N. J. (2008) *Creating scientific concepts.* The MIT Press.

Owensby, J. N. (2006). *Exploring the development and transfer of case use skills in middle school project-based inquiry classrooms* [Unpublished doctoral dissertation]. Georgia Institute of Technology.

Papert, S. (1993). *The children's machine: Rethinking school in the age of the computer.* Basic Books.

Pólya, G. (1945). *How to solve it: A new aspect of mathematical method.* Princeton University Press.

Repenning, A., & Ioannidou, A. (2008). Broadening participation through scalable game design. *ACM SIGCSE Bulletin, 40*(1), 305–309.

Repenning, A., Webb, D., & Ioannidou, A. (2010). Scalable game design and the development of a checklist for getting computational thinking into public schools. *SIGCSE.* https://doi.org/10.1145/1734263.1734357

Roschelle, J. (1996). Learning by collaborating: Convergent conceptual change. In T. Koschmann (Ed.), *CSCL: Theory and practice of an emerging paradigm* (pp. 209–248). Lawrence Erlbaum Associates.

Schneider, G. M., & Gersting, J. (2019). *Invitation to computer science* (8th ed.). Cengage.

Schoenfeld, A. H. (2010). *How we think: A theory of goal-oriented decision making and its educational applications.* Routledge.

Thomas, J. O. (2008). *Scaffolding complex cognitive skill development: Exploring the development and transfer of case use skills in middle school project-based inquiry classrooms.* VDM Publishing.

Thomas, J. O. (2014). Supporting Computational Algorithmic Thinking (SCAT): Exploring the development of computational algorithmic thinking capabilities in African-American middle school girls. In J. L. Polman, E. A. Kyza, D. K. O'Neill, I. Tabak, W. R. Penuel, A. S. Jurow, K. O'Connor, T. Lee, & L. D'Amico (Eds.), *Learning and becoming in practice,* (Vol. 2, pp. 1092–1096). International Society of the Learning Sciences.

Thomas, J. O. (2015, June 22–25). *Supporting computational algorithmic thinking (SCAT): Understanding the development of computational algorithmic thinking capabilities in African-American middle school girls through game design* [Paper presentation]. Tenth International Conference on the Foundations of Digital Games, Pacific Grove, CA, United States. http://www.fdg2015.org/papers/fdg2015_paper_17.pdf

Thomas, J. O. (2018). The computational algorithmic thinking (CAT) capability flow: An approach to articulating CAT capabilities over time in African-American middle school girls. In T. Barnes & D. Garcia (Eds.) *In Proceedings of the 49th ACM Technical Symposium on Computer Science Education* (pp. 149–154). ACM SIGCSE.

Thomas, J. O., Minor, R., & Odemwingie, O. (2016). Exploring African-American middle school girls' perceptions of computational algorithmic thinking of themselves as game designers. In C. K. Looi, J. L. Polman, U. Cress, & P. Reimann (Eds.), *Transforming learning, empowering learners,* (Vol. 1, pp. 960–962). International Society of the Learning Sciences.

Thomas, J. O., Minor, R., & Odemwingie, O. C. (2017a). Exploring African-American middle school girls' perceptions of themselves as game designers. In Y. Rankin & J. Thomas (Eds.), *Moving students of color from consumers to producers of technology* (pp. 49–61). IGI Global. https://doi.org/10.4018/978-1-5225-2005-4

Thomas, J. O., Odemwingie, O. C., Saunders, Q., & Watlerd, M. (2015). Understanding the difficulties African-American middle school girls face while enacting computational algorithmic thinking in the context of game design. *Journal of Computer Science and Information Technology, 3*(1), 15–33.

Thomas, J. O., Rankin, Y., Minor, R., & Sun, L. (2017b). Exploring the difficulties African-American middle school girls face enacting computational algorithmic thinking over three years while designing games for social change. *Computer Supported Cooperative Work, 26*(4–6), 389–421.

Vygotsky, L. S. (1978) *Mind and society: The development of higher mental processes.* Harvard University Press.

Wells, G., & Chang-Wells, G. L. (1992). *Constructing knowledge together: Classrooms as centers of inquiry and literacy.* Heinemann.

Werner, L. L., Campe, S., & Denner, J. (2005, October). Middle school girls + games programming = information technology fluency. In *Proceedings of the 6th Conference on Information Technology Education* (pp. 301–305). ACM. https://doi. org/10.1145/1095714.1095784

Werner, L., Denner, J., Bliesner, M., & Rex, P. (2009, April). *Can middle-schoolers use Storytelling Alice to make games? Results of a pilot study* [Paper presentation]. Fourth International Conference on the Foundations of Digital Games, Orlando, FL, US.

Wing, J. (2008). Computational thinking and thinking about computing. *Philosophical Transactions of the Royal Society A: Mathematical, Physical and Engineering Sciences, 366*(1881), 3717–3725. https://doi.org/10.1098/rsta.2008.0118

Wing, J. M. (2006). Computational thinking. *Communications of the ACM, 49*(3), 33–35. https://doi.org/10.1145/1118178.1118215

Wing, J. M. (2010). *Computational thinking: What and why?* [Unpublished manuscript]. Computer Science Department, Carnegie Mellon University. https://www.cs.cmu. edu/~CompThink/resources/TheLinkWing.pdf

Yadav, A., Zhou, N., Mayfield, C., Hambrusch, S., & Korb, J. T. (2011, March 9–12). *Introducing computational thinking in education courses.* In *Proceedings of the 42nd ACM Technical Symposium on Computer Science Education* (pp. 465–470). SIGCSE. https:// cs4edu.cs.purdue.edu/_media/ctmods_sigcse11.pdf

Zweben, S., & Bizot, B. (2017). Taulbee Survey: Generation CS continues to produce record undergrad enrollment; Graduate degree production rises at both master's and doctoral levels. *Computing Research News, 29*(5), 3–51.

3

CODING, GAME DESIGN, AND COMPUTATIONAL THINKING

Jacqueline Leonard

I love playing The Sims. I recall purchasing the original game in 2002. At that time, I created a family of four—husband, wife, and two children. After creating the family, I could choose different personalities and characteristics for each family member, such as materialistic, perfectionist, shy, friendly, etc. Then I created their physical features and clothing. After that, I created a home for the family to live in. I chose a three-bedroom, one-story home with upgrades based on the funds they had available. The husband had a job in the business industry, and the wife worked in politics. They had a baby and a school-aged child to care for. I made sure that the parents made it to work on time and the child was outside to catch the school bus. The mother had to make a phone call to hire a babysitter for the infant so that she could go to work. In order to be promoted on their job, the adults had to improve on certain traits or skills, such as writing, charisma, socializing, making friends, etc. The school-aged child could work hard or be average at school and needed to complete homework in order to improve his/her grade. When the family returned from work, I had to address their basic needs, such as hygiene, nutrition, exercise, socialization, and leisure. The baby had to be fed and diapers had to be changed often to model good care and nurturing. If the baby cried too much, as a result of hunger or poor hygiene, a social worker might come and take the baby away. The power to create characters, choose their careers, and manage their lives was intriguing. The game also allowed me to establish genealogy by creating snapshots of parents and grandparents to show lineage. As the original Sims evolved to other versions, characters were able to own and manage businesses, go to college, go on vacations, or own pets. I have purchased every expansion pack. The real-life aspect of the game, along with the genealogy, and having choices or options kept me interested in the game for almost 20 years.

The impetus for Jacqueline's (first author's) research interest in gaming and game design is three-fold. First, the vignette given earlier described her daughter's, Mia's (pseudonym), experiences with *The Sims*. More than a preoc-cupation, *The Sims* hooked Mia into becoming a lifelong gamer. She played *The Sims* for hours on end and became annoyed when an interruption made her lose focus on the task at hand. This observation led Jacqueline to believe other children could be genuinely excited about gaming and game design as well. Second, Jacqueline recalled how her students enjoyed LOGO during her teaching experiences in the 1980s. At that time, LOGO involved manipulat-ing a turtle on a green screen by typing simple codes like LT 30 or RT 45. Students were able to learn geometry while having fun rotating and moving the turtle. Third, Jacqueline was influenced by the work of Alexander Repenning, a colleague at the University of Colorado Boulder. She was invited to take a summer course on game design. *Scalable Game Design* (SGD) allows users to employ drag-and-drop features and to input simple if-then codes that execute commands for agents (i.e., avatars) to move within the background of a game or simulation (Repenning et al., 2015, 2017). SGD provides scaffolding as well as multiple opportunities for creativity as the background and agents can be whatever the user wants them to be.

After a brief review of the extant literature, digital gaming and game design contexts will be discussed in terms of their educational value, espe-cially as it relates to females—a population that is often underrepresented in STEM in general and technology specifically. Games can also be used as con-texts for Black and other marginalized students to express themselves and/ or ameliorate how they are treated in society at large (Gholson & Robinson, 2019; Leonard & Hill, 2008). Interventions that embed opportunities for stu-dents of color to address the violence they experience in their communities is restorative (Gholson & Robinson, 2019) and redemptive (Leonard, 2009). Culturally based research and its applicability to classroom practices will also be discussed with the aim of helping K-8 teachers, teacher educators, and STEM educators understand how digital gaming can be used to improve mathematics achievement, which remains a gatekeeper to academic success in STEM (Martin et al., 2010).

Educational Use of Gaming

In the 1980s and 1990s, computer-based games like *Reader Rabbit* and *Math Blaster* were used in most K-8 classrooms primarily as remedial tools. Other games such as the *Oregon Trail* provided students with decision-making options and opportunities to engage in problem-solving skills. These and other computer-based games provide students with immediate feedback aligned with literacy and learning goals (Renaud & Wagoner, 2011). Although feedback is a positive aspect of these computer-based programs, the focus on learning and practicing basic skills in individualized programs isolated students from their peers (Kitchen & Beck, 2016). Since 2010, students have been able to engage in games that mimic real-life experiences (i.e., *The Sims*), design games and simulations that allow them to be creators rather than consumers (i.e., *Scratch, Microworlds EX, Scalable Game Design*) (Leonard, 2019), and participate in gaming communities socially with other children (Gee & Hayes, 2010; Kafai & Burke, 2014).

While the learning curve to play or produce such games may be high, games should appeal to educators because the "most-engaged learning is often right on the edge of being too difficult" (Renaud & Wagoner, 2011, p. 59). Koutromanos's and Avraamidou's (2014) review of the literature found that mobile games and apps helped students to develop positive attitudes toward mathematics. Moreover, Chang et al. (2012) found that digital game playing supported the development of critical thinking and mathematics problem-solving skills. Gaming not only supports systematic thinking but allows students to connect ideas and use multiple strategies to solve problems. Strategic thinking and systematic thinking to solve problems are examples of computational thinking (Newton et al., 2020; Renaud & Wagoner, 2011; Sullivan & Heffernan, 2016). Thus, coding and game design can be used to engage students in computational thinking and has the potential to increase mathematics outcomes (Ritzhaupt et al., 2011).

The Sims

According to Gee and Hayes (2010), *The Sims*, along with all of the expansion packs, is one of the best-selling videogames in history. *The Sims* provides learners with a context to create a virtual family where family members go through the challenges of everyday life, such as getting a job, earning wages, and paying bills. While gamers can spend hours playing *The Sims*, they can also spend time designing content. Thus, a number of communities have been formed to support acquisition of 21st-century skills and novel learning communities through the Internet (Gee & Hayes, 2010). The communities that have emerged around *The Sims* primarily consist of girls and women, who are often marginalized in STEM and gaming, but "central to where gaming, popular culture, and learning are going in the future" (Gee & Hayes, 2010, p. 16). While some may consider *The Sims* a glorified dollhouse, female gamers learn 21st-century skills along with emotional intelligence (i.e., ability to manage sentiments, relationships, viewpoints, and social interactions) (Gee & Hayes, 2010). Moreover, *The Sims* can be used to enhance literacy skills, particularly among English learners (ELs) (Renalli, 2008).

The real-life context of *The Sims* provided ELs with opportunities to improve their grammar and vocabulary through gaming (Ranalli, 2008). Using mixed methods, Ranalli (2008) replicated a study that used simulations—which, like digital games (Li, 2010), can "foster strategic and communicative competence by helping learners assess the characteristics of a language-use situation" to enhance language learning (Renalli, 2008, p. 442). Furthermore, the gaming exposed students to written language, such as instructions and labels. Renalli's study consisted of nine undergraduates who were classified as English as a Second Language (ESL) students. Thirty vocabulary words that were associated with *The Sims* were selected for the intervention and presented in charts along with definitions, synonyms, antonyms, and contextual meanings. Supplemental materials served as a companion to the game as well as a means to administer the treatment at three different stations. Pre-post measures were used to assess knowledge of the 30 vocabulary words. Results suggest that combining the supplemental materials with playing *The Sims* contributed to vocabulary acquisition. Thus, in addition to problem-solving and 21st-century skills, gaming was used to help ELs to develop literacy skills.

LOGO

LOGO (i.e., logic oriented graphic oriented) is a programming language that was developed in the 1960s by Seymour Papert and colleagues to investigate the notion that children could learn mathematics and engage in metacognition through programming (Kafai & Burke, 2014; Manches & Plowman, 2017). Children were able to write code by using simple commands to move what became known as an "on-screen turtle graphic" to learn logical problem-solving skills (Manches & Plowman, 2017, p. 195). The idea of using logic to solve problems was later defined as algorithmic problem solving (Barr & Stephenson, 2011), which is akin to computational thinking. Papert is known as the father of computational thinking while Wing (2006) was one of the first to define it but revised the definition to include "the thought processes involved in formulating problems and solutions so that the solutions are represented in a form that can be effectively carried out by an information-processing agent" (Wing, 2010, p. 1). Thus, problem solving is not only essential in computational thinking but also in formulating the problems to be solved. Such problems can be created within the advanced web-based version of LOGO known as *Microworlds EX*, which is a computer programming tool that relies on graphic user interfaces (GUIs) to develop computer simulations. In *Microworlds EX*, the on-screen turtle graphic may be any object.

In a 2015 study, *Microworlds EX* was used as the intervention to examine 17 Black and Latinx students' self-efficacy in technology, 21st-century skills, STEM attitudes, and attitudes toward STEM careers during a summer STEM camp (Leonard et al., 2019). Students created simulations that ranged from horse racing to downhill skiing. Latinx students also created culturally based images to illustrate life in their communities. Because of the small sample size, a non-parametric test was performed to determine whether the study yielded any significant differences from pre- to posttest on the survey items. Using a confidence interval of 0.90, the results revealed significant differences on the constructs of 21st-century skills ($p = 0.026$), science attitude ($p = 0.036$), and engineering careers ($p = 0.058$). The effect sizes were moderate. While mathematics and engineering attitudes did not increase significantly, these scores trended upward and remained stable. Thus, computer simulations like *Microworlds EX* can be used to engage underrepresented children in computational thinking and cultural relevance to broaden their opportunities to participate in STEM (Leonard et al., 2019).

Scalable Game Design

Scalable Game Design (SGD) was developed by Alexander Repenning and colleagues at the University of Colorado Boulder. The researchers identified several computational thinking patterns (CTPs) that students used while learning to code within the SGD platform (Repenning et al., 2017). CTPs consist of absorption, generation, collision, transportation, diffusion, polling, and hill climbing, which provide learners with a floor to acquire initial understanding of digital skills (Jenson et al., 2016; Li, 2010). Repenning et al. (2017) found that SGD not only provided the scaffold for students to develop agents but also opportunities for them to engage in critical thinking skills as they used if-then reasoning to design games. The drag-and-drop features along with conversational programming allowed students to engage in three functional stages of computational thinking: problem formulation (i.e., abstraction), solution expression (i.e., automation), and execution (i.e., analysis) (Repenning et al., 2017).

Because of the robust nature of the SGD programming language and the flexibility students had to design the background and agents, Jacqueline (first author) and her colleagues used SGD to examine computational thinking in a research study that consisted of 807 student participants over a three-year period. Several aspects of the study, known as uGame-iCompute (UGIC), have been reported elsewhere (Leonard, 2019; Leonard et al., 2016, 2018; Newton et al., 2020). However, none of these reports focused exclusively on computational thinking and game design. A synopsis of the research setting and findings from the Year 2 study that consisted of two afterschool programs and a summer camp are presented.

The uGame-iCompute Study

Third- through sixth-grade students participated in the UGIC study during informal afterschool programs from spring 2014 to summer 2016. Most of the students were White and attended rural or small urban schools in the state of Wyoming. During the pilot study, student participants were offered opportunities to learn robotics and/or game design. In Year 2, the researchers isolated the treatment variables. In the fall of 2014, students engaged in robotics only, and in spring of 2015, they engaged in game design. In Year 3, students participated in both robotics and game design—20 hours each—for a total of 40 hours of intervention. Findings from Year 2, as they relate to game design, are reported in this chapter. Data were collected from three different sites that included the

University Lab School (Site 1), a small rural school in Northwest Wyoming (Site 2), and a summer camp held at a community-based site in the state capital (Site 3). Students at these sites represent convenience samples that were selected because they provide nuanced perspectives on game design that collectively show impact on student outcomes and computational thinking.

The research questions that guided the study reported here were as follows:

1. What was the relationship between the number of hours students spent coding and their growth in mathematics achievement?
2. What CTPs emerged among focal students on game design and how did they compare and contrast during an informal afterschool program?
3. How did focal students' computational thinking skills on game design compare and contrast during a summer STEM camp?
4. How did students' attitudes toward engineering/technology and 21st-century skills change after participating in a summer STEM camp?
5. How did focal students describe their perspectives on game design after participating in a summer STEM camp?

Method

"Case studies are detailed examinations of singular settings, subjects, or events" that "allow one to gain deeper understanding of local experiences that other research designs do not offer" (Ryoo, 2019, p. 36:5). The case study method was used to provide an ethnographic view of the learning that took place during two afterschool programs and one summer camp where the principal investigator and first author of this text was a participant observer. Mixed methods were used to analyze the data collected in the Year 2 study. Quantitative methods were used to analyze survey data. Qualitative methods consisted of analyzing field notes and transcribing 45-minute to 60-minute lessons on students' engagement in game design.

The Case Study Participants

With consent from the Institutional Review Board and support from school administrators and 25 teacher participants, upper elementary and middle school students were recruited from 17 schools in Year 2. Data collected on game design from three of these sites are presented to examine students' CTPs and skills. Case study 1 reports on student participants who attended an afterschool program at the K-9 University Lab School (Site 1, $n = 38$). These students participated in MINDSTORMS® coding for LEGO® robotics in fall 2014 and game design in spring 2015. Case study 2 reports on student participants who attended an afterschool program at a rural K-6 school (Site 2, $n = 11$) during spring 2015. These students focused on game design only. Case study 3 reports on students recruited in the state capital who attended a community-based program in summer 2015 that focused on robotics and game design. However, only outcomes from the game design course are reported for student participants at these sites. Demographics for the combined student population were: 70% male and 30% female; 84.3% White, 5.7% Asian, 5.7% Hispanic, 2.9% Native American, and 1.4% Black/African American.

Procedures

Prior to conducting the project, the teachers received professional development by participating in an online course that was delivered by project staff with expertise in computer science. Female teacher facilitators in the first two case studies taught SGD during afterschool programs. Game design was taught by the principal investigator during the summer STEM camp, soon after she completed the training course on SGD at the University of Colorado in Boulder.

Students participated in an average of 20 hours of game design during afterschool or summer camp programming in Year 2. In order to design the games, students downloaded SGD software on laptop computers. They were immediately presented with a white screen and a pallet that allowed them to make a background for the game. For the background, students could draw and repeat a pattern or simply use a single color from the pallet. After creating the background, students produced the agents for the game. Agents could be objects and characters that were able to move on the game board. An example of one female participant's Frogger game is shown in Figure 3.1. The agents were programmed using drag-and-drop codes to specify a number of behaviors including moving in various directions (e.g., up, down, left, right), producing sound effects, and restarting or ending the game if an agent died. Within the game, agents could carry (i.e., transport) other agents or follow a scent (i.e., diffusion) to chase another agent (Repenning et al., 2017). Students could also perform more sophisticated

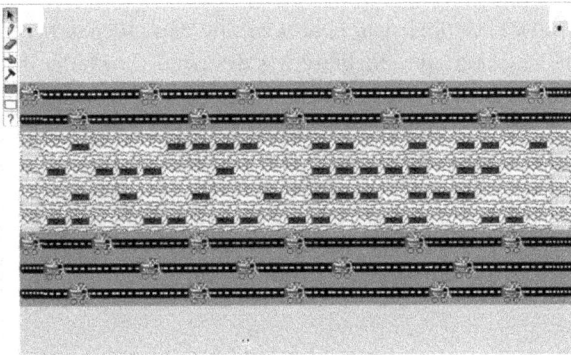

FIGURE 3.1   Screenshot of Sample Frogger Game

steps, known as hill climbing, to increment the difficulty of a game (Repenning et al., 2017). Thus, students had a low floor (i.e., minimal skill entry level), high ceiling (i.e., ability to engage in advanced gaming), and wide walls (i.e., ability to be creative and expressive) (Kafai & Burke, 2014) to develop authentic games to demonstrate computational thinking.

Data Sources and Analyses

Quantitative data consisted of students' Measures of Academic Progress (MAP) scores in mathematics and student attitudes toward engineering/technology and 21st-century skills (Cronbach alpha ≥ 0.83) (Unfried et al., 2015). The Statistical Package for the Social Sciences (SPSS, Version 26) was used to analyze the relationship between time and mathematics growth at Site 1 using Pearson's correlation coefficient. In order to account for outliers (range from 6 to 85 hours), students who participated in less than 20 hours were placed into one group, and students who participated in 20 (median number) or more hours were placed into a second group. To determine if there was a significant difference between the number of coding hours and mathematics growth, the one-way Analysis of Covariance (ANCOVA) was used. Additionally, students in the summer camp completed a modified survey to ascertain student attitudes toward engineering/ technology and 21st-century skills. The *T-statistic* was used to compare pre-post scores by conducting a paired *t*-test.

Qualitative data consisted of field notes on teachers' lesson plans and students' activities collected by the researchers. The processes that students went through to develop the games were obtained from these data sources. At Site 2, focal students' games were analyzed qualitatively for elements of CTPs. The principal investigator developed a rubric to characterize computational thinking based on the International Society for Technology in Education (ISTE, 2020) framework (e.g., abstraction, generalizing, and transfer, etc.), which was piloted at Site 3. The three-point rubric rated student games as emerging (1), moderate (2), or substantive (3) based on evidence of computational thinking across six constructs (see Appendix A). The rubric was field-tested by two independent raters on the research team, and interrater reliability was determined to be 86%. Finally, focal students at Site 3 also participated in small group interviews to examine their perspectives on using game design as a tool for STEM engagement. The interviews were transcribed verbatim, read, and analyzed for patterns using the constant comparative method (Glaser, 1965). Student responses provided additional evidence to support survey data.

Results

The results of the three case studies are reported for each site. Data collected and analyzed at Site 1 show the potential impact of game design on student achievement outcomes. Data collected and analyzed at Site 2 and Site 3 provide two different methods for assessing students' computational thinking. Data collected and analyzed at Site 3 provide insight into student perspectives about game design.

MAP Scores

To answer the first research question about the relationship between the hours of coding and students' mathematics growth, we analyzed student data at Site 1. The Pearson correlation coefficient was used to determine the relationship between the number of hours spent coding and students' growth scores after participating in the study. Ms. Adams (pseudonym), one of the focal teachers at Site 1, shared these data with the research team at the conclusion of the Year 2 study. The results of the Pearson Product-Moment Correlation revealed a moderate correlation ($r = 0.441$, one-tailed) between the number of hours students ($n = 25$) spent coding and their growth on the MAP mathematics test from fall 2014 to spring 2015. This correlation was significant ($p = 0.011$) at a confidence interval of 0.95. In order to compare student growth by hours of coding on the MAP mathematics assessment, the one-way ANCOVA was performed using the growth from 2012 to 2013 (i.e., year of no intervention) as a covariate. As shown in Table 3.1, results of the ANCOVA comparing growth by hours of coding revealed a significant main effect when students participated in 20 or more hours of coding: $F(1, 21) = 8.743, p = 0.008$. Partial eta squared was 0.315, which shows a large effect size. Nevertheless, given the small sample size, these results should be interpreted with caution.

Computational Thinking Patterns and Game Design

According to Repenning et al. (2017), CTPs provide students with a robust high-level language to think about problem solving before they begin to code. CTPs are evident when students use code to program behaviors that allow the agents in the game to perform one or more of seven actions: generation, absorption, collision, transportation, diffusion, polling (i.e., act of counting to determine if the game is won), and hill climbing (i.e., advancing from one level to another to increase difficulty).

TABLE 3.1 MAP Mathematics Growth (2015) by Hours of Coding Controlling for Treatment

Hours of Coding	Mean MAP Growth	SD
Group 1 (19 hrs or less)	4.25 (n=8)	7.09
Group 2 (20 hrs or more)	11.71 (n=14)	6.01

Note: Measures of Academic Progress

To answer the second research question about the emergence and comparative aspects of CTPs, lesson plans and student artifacts at a rural school (Site 2) were analyzed. The lessons taught by Ms. Cobb at the rural school (see Leonard et al., 2018) provided a context for analyzing focal students' artifacts in great detail. Ms. Cobb was observed facilitating student learning on two different occasions in the month of March 2015. She often began by placing directions on a screen using a Promethean board. While most of the teacher facilitators used the SGD tutorial known as AgentSheets to instruct the students, Ms. Cobb developed EDMOTO instructions for the students to follow. Additionally, she often provided one-to-one support as needed.

During the second observation, several students worked independently on their game designs. Screenshots of the work revealed some of the CTPs that occurred during this session. While every behavior may not be present in every game, the use of multiple behaviors (CTPs) and understanding how they work show the depth of students' computational thinking in terms that demonstrate their ability to solve problems using a systematic process. Seven of the 11 student participants' games were analyzed for CTPs. Three games were mazes, and four were Frogger games. Analyses of the games revealed information about specific CTPs. For example, as agents collided in the games, sound effects (e.g., kaboom, gunshots, etc.) or text messages (e.g., "You win!"; "You die!") were programmed into the game to alert the user that the game had ended or reset. Table 3.2 shows analyses of students' games by the type and number of CTPs. These games did not allow for polling or diffusion, and beginning students are generally not advanced enough to engage in hill climbing. Therefore, polling, diffusion, and hill climbing are not included in the analyses.

Focal students engaged in four unique CTPs for a total of 44 instances, which was about six instances per game. On average, students who produced maze games had higher instances of behaviors associated with collision. However, resetting the game without embedding the code to stop the game results in an infinite loop. This was the case with student 3-315's game. The Frogger game allowed students to use the most CTPs in this case study. Transportation was

TABLE 3.2 Analyses of CTPs in Rural Focal Students' Game Designs

Student#	Type	Absorb	Generate	Collide	Transport	Notes
1-315	maze		1	5		
2-315	maze		1	3		Game was in beginning stage
3-315	maze		1	6		Recorded steps; infinite loop
4-315	Frogger	2	2	2	1	
5-315	Frogger	2	2		1	Game was in beginning stage
6-315	Frogger	3	3	1	1	Recorded steps
7-315	Frogger	2	2	2	1	Recorded steps
Totals		9	12	19	4	

utilized the least but was evident in each of the Frogger games. Three of the seven focal students explained the coding to show their understanding about how the program worked. Figure 3.2 shows student 6-315's code for a Frogger game with an explanation of each step. Coding is explained as follows: *If I see water then I will*

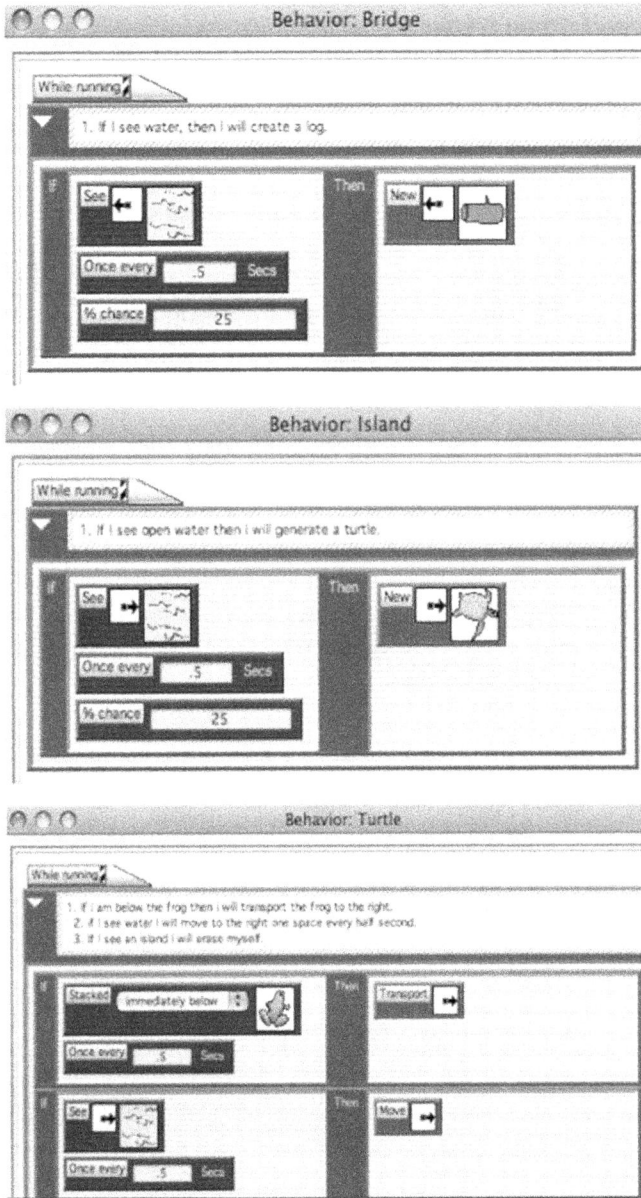

FIGURE 3.2 Sample Code with Explanations

create a log (i.e., generation); *If I see an island I will erase myself* (i.e., game is over); *If I am below the frog then I will transport the frog to the right* (i.e., transportation). Overall, students used CTPs effectively to develop digital games.

ISTE Computational Thinking Skills and Game Design

To answer the third research question, which focused on comparing and contrasting CTPs in a STEM context, data collected during a one-week summer STEM camp were analyzed at Site 3. The principal investigator taught the game design course to 24 students (19 boys and 5 girls) who were recruited to participate in a community-based program in August 2015. Students developed three types of games—mazes, Frogger, and Pac-Man. These games allowed students to engage in six of the seven CTPs, including diffusion and polling but not hill climbing. The principal investigator developed a rubric to rate students' computational thinking more broadly using the ISTE standards (ISTE, 2020). The rubric was developed to use with game designs that were not produced using the SGD platform. Additional games, such as *Scratch*, were used in Year 3 of the study (Leonard, 2019). Thus a more general rating scale was needed to assess students' computational thinking regardless of the platform. Twelve focal students were randomly selected for analysis. Their games were rated by a graduate student, who had been trained to use the rubric, for evidence of CT. Four of the games were mazes, four were Frogger games, and four were Pac-Man games to ensure that each type of game was adequately represented in the data analysis. Students' games were evaluated based on coding and game functionality. Students' scores are reported in each of the ISTE categories as well as by average individual and total mean scores (see Table 3.3).

Analyses of data in Table 3.3 reveal that focal students' computational thinking improved as they learned how to produce additional types of games. The maze game was taught first, followed by Frogger and then Pac-Man. Recall that scores were rated using a 3-point rubric: emerging (1), moderate (2), and substantive (3). While patterns varied by the type of game, most students appeared to score highest on abstraction and lowest, with some exceptions, on algorithms, analyzing/implementing or generalizing/transfer, which are more sophisticated skills. In order to score well on generalization/transfer, students had to demonstrate that they could deviate from the AgentSheets tutorial, which was used initially to learn the basic principles of SGD. The data revealed that the Frogger game was more difficult to deviate from than maze and Pac-Man. Table 3.4 summarizes the students' scores based on the rubric.

Table 3.4 shows that two maze games were rated as emerging. Two mazes and three Frogger games were rated as moderate. All four Pac-Man games were rated as substantive. Overall, the total mean score was highest on Pac-Man ($M = 2.67$), which was a more difficult game to produce than Frogger ($M = 2.21$) or mazes ($M = 1.46$). The implication is that students got better as they progressed

TABLE 3.3 Analyses of Computational Thinking Skills (ISTE Standards)

Student#	Problem Formulation	Abstraction	Logical Thinking	Algorithms	Analyzing & Implementing	Generalizing & Transfer	Mean CT Ratings
Maze							
1–815	1	2	1	1	1	1	1.17
2–815	1	2	1	1	1	1	1.17
3–815	2	3	2	1	1	2	1.83
4–815	1	3	1	1	1	3	1.67
Mean Score	**1.25**	**2.50**	**1.25**	**1.00**	**1.00**	**1.75**	**1.46**
Frogger							
5–815	2	3	2	2	1	1	**1.83**
6–815	3	2	3	3	3	1	2.50
7–815	3	3	1	3	2	1	2.17
8–815	3	2	3	3	2	1	2.33
Mean Score	**2.75**	**2.50**	**2.25**	**2.75**	**2.00**	**1.00**	**2.21**
Pac–Man							
9–815	3	3	2	2	2	3	**2.50**
10–815	3	2	3	3	3	1	2.50
11–815	3	3	3	3	2	3	**2.83**
12–815	3	3	3	2	3	3	2.83
Mean Score	**3.0**	**2.75**	**2.75**	**2.50**	**2.50**	**2.50**	**2.67**

TABLE 3.4 Summary of Focal Students' CT Rating by Game Type

	Emerging (M = 1–1.4)	Moderate (M = 1.5–2.4)	Substantive (M = 2.5–3.0)
Maze	2	2	
Frogger		3	1
Pac-Man			4

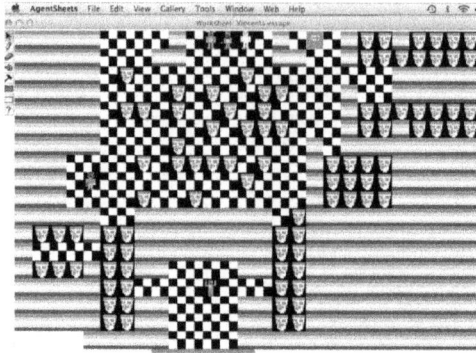

FIGURE 3.3 Sample Pac-Man Game

throughout the course. Thus, students' computational thinking skills improved over time as focal students demonstrated greater ability to design functional games near the end of the course. This indicated a greater than average ability to engage in problem formulation, abstraction, logical thinking, algorithms, analyzing/implementing, and generalizing/transfer. A sample Pac-Man game is shared as an example of these elements (see Figure 3.3).

The Pac-Man game was created by a female (student 9.815) with Asperger's syndrome. She called the game Foxy Run, which is based on an off-shelf videogame called *Five Nights at Freddy's*. This student, as well as some others, embedded elements (e.g., agents and sound effects) of pop culture into her game. The abstract nature of the game as well as its functionality resulted in an overall score of 2.50, which demonstrated substantive CT on the three-point rubric. Only one out of five students in case study 3 was female. Female CT scores ($n = 3$) are shown in bold text (see Table 3.3). Like their male counterparts, females' CT scores tended to increase as they progressed throughout the course. There were too few female students in the course to draw any direct conclusions about their gaming ability. However, field notes indicated that the two females who exhibited substantive CT scores considered themselves to be gamers.

Learning from Student Focus Groups

In order to answer the fourth and fifth research questions, quantitative data collected from surveys of students' attitude toward STEM and qualitative data collected during focus group interviews were analyzed at Site 3. The results of a paired *t*-test revealed no significant differences among this cohort of students (n = 21) on pre-post surveys of attitudes toward engineering/technology (*M*pre = 3.87 [*SD* = 0.80]; *M*post = 3.93 [*SD* = 0.72]) and 21st-century skills (*M*pre = 4.12 [*SD* = 0.72]; *M*post = 4.24 [*SD* = 0.65]). However, qualitative data allowed individual student voices to be heard. The focus group interview was conducted by the project manager on the research team. The responses of three females and four males as it related to queries about game design are presented here.

INTERVIEWER: **So now, I'm going to ask you the same questions, but now I just want you to tell me about the gaming part. What one or two things do you think was/were really cool about gaming?**

I think the coolest thing about the gaming was how you got to make your own avatar things out of the pixels and also, the programming was fun, too. (Female 1)

I liked making Frogger and being able to play it. Right now, my favorite game is my first game. It's something like Pac-Man, and I made Snakes. It's really fun! (Female 2)

What I like about the gaming is that we get to try new things and we get to make games out of [inaudible] and make it our own, and I like that we get to make new things and try it out. (Male 1)

I liked how you could make it impossible for people not to beat your game, because that's what I did, and that our teacher... told me that she'd never tried this [if-then condition] before. (Male 2)

I liked creating my own game, just the creativity of it, and also, you could actually play the game and see things move. (Male 3)

INTERVIEWER: **For how many of you is this the first time you've even had an opportunity to program games?**

All participants raise hands.

INTERVIEWER: **What didn't you like?**

I don't really like doing gaming because I like putting things together, and it was so hard for me because the computer did not work properly. (Female 3)

I didn't really like the gaming either because... (Interviewer: you preferred to put stuff together?) *...Yeah.* (Male 4)

I think it wasn't exactly the program; it was more the computer because the mouse would mess up and then it would pull out of the program, and it was just really hard sometimes. (Male 1)

What I didn't like about it was that it's not as easy as robotics because with gaming you have to follow specific instructions, and it was hard because your program could get really easily confusing and you could do something wrong and it was just stressful when you do something wrong. (Male 4)

INTERVIEWER: **Did you know you could upload your game to the Internet and people can play them? Oh you're going to learn that? Well is that something that you'd want to do once you finish your game?**

Yeah! (All participants)
Except that I'd want to create a different kind (of game). (Male 2)

Using the constant comparative method (Glaser, 1965) to analyze the focus group interview data, it is clear that the majority of the students (i.e., two females and three males) enjoyed participating in game design. Themes and patterns that emerged from the data include the following: (a) creativity or making games; (b) game playing; (c) programming; (d) novelty of game design; (e) glitches/difficulty; and (f) lack of autonomy. Five students made comments about enjoying the creativity associated with game design. Some comments included the following: *I think the coolest thing about the gaming was how you got to make your own avatar things out of the pixels; I liked creating my own game.* Three students made positive comments about playing the games: *I liked making a Frogger and being able to play it; …You could actually play the game and see things move; I like that we get to make new things and try it out.* Two students enjoyed the programming: *…the programming was fun, too; I liked how you could make it impossible for people not to beat your game.* All students mentioned the novelty of game design and the fact that the summer STEM course created their first opportunity to participate in this kind of activity. For some, game design created a deep learning curve that caused two students to have negative attitudes: *I don't really like doing gaming because I like putting things together, and it was so hard for me….; …it was hard because your program could get really easily confusing.* These comments were coupled with two comments about glitches causing the program to freeze or work improperly: *…it was more the computer because the mouse would mess up and then it would pull out of the program, and it was just really hard sometimes; … it was so hard for me because the computer did not work properly.* Finally, two students mentioned lack of autonomy in terms of the kinds of games the SGD platform allowed them to make: *I'd want to create a different kind; with gaming you have to follow specific instructions.* This critique is important, as it suggests that historically-excluded students benefit from software tools and platforms that support developing games that illustrate unique aspects of their cultures and communities (Leonard & Hill, 2008). In summary, the majority of students enjoyed game design, and the interview data informed the researchers that students should have greater choices that allow them to make connections to their cultural identity (Gholson & Robinson, 2019).

Discussion

Several findings emerged from the results of the Year 2 UGIC reported in this chapter. The first finding is the number of hours one sample of students spent coding was related to mathematics outcomes on the MAP test. While the range of hours in the case study at Site 1 varied greatly, growth scores increased significantly if students spent more than 20 hours coding. This finding concurs with the research findings of Ritzhaupt et al. (2011) that digital games increased students' mathematics achievement. However, it also challenges teachers, teacher educators, and researchers to find ways to encourage all students to participate in opportunities to code. While expanding the menu of game design programs to include *Scratch* and off-shelf games like *The Sims* is one way to broaden participation, the software should allow for student autonomy and self-expression. Students should to be able to see themselves as gamers by creating apps of their own (Gee & Hayes, 2010).

The second finding is that students at Site 2 effectively used CTPs and students at Site 3 incrementally improved their use of computational thinking skills as defined by the ISTE Standards (2020). At Site 2, students used an average of two CTPs to develop maze games while three or four CTPs were used to develop Frogger games. At Site 3, students' maze games were rated emergent (i.e., beginner level) while more advanced games ranged from moderate to substantive in terms of CT. Thus, students exhibited more robust CTPs (Site 2) and computational thinking skills (Site 3) as they gained more knowledge and experience with SGD. This was true for female as well as male focal students at Site 3. This finding is encouraging and concurs with the finding that females are avid gamers (Gee & Hayes, 2010).

The third finding is that students who have a pre-existing interest in STEM did not suffer or lose interest as a result of participating in coding activities like MINDSTORMS® and SGD. While the pre-post scores on a survey that measured engineering/technology attitudes and 21st-century skills (Unfried et al., 2015) did not increase significantly, scores did not significantly decline either. This result differed from the pilot study where engineering/technology scores declined (see Leonard et al., 2016) and from the Year 3 study where engineering/technology and 21st-century skills scores also declined (Leonard, 2019). Yet, the results of the case study at Site 3 should be interpreted cautiously because of the small sample size. Perhaps these findings were the result of ceiling effects as students' initial scores approached 4 on the survey's 5-point Likert scale. Nevertheless, these results warrant further research where students engage in game design exclusively and not in combination with robotics.

The fourth finding reveals the majority of focal group students had positive views of game design. Two female students in particular mentioned the ability to create agents as the primary reason for their positive perspectives on game design. These comments are consistent with previous findings on digital gaming and student motivation (Koutromanos & Avraamidou, 2014). Their comments also align with Mia's remarks in the introductory vignette. She stated that she enjoyed creating and managing the lives of characters that she developed in *The Sims*. Because of the sparse number of girls participating in the case study at Site 3, the three females in the focus group were the same girls whose work was showcased in Table 3.3. The girls who identified as gamers and had substantive ratings on the Pac-Man game were the same two girls that had positive perspectives about game design. The third female preferred to build things like robots and thus preferred pre-engineering tasks. Therefore, it is important to support STEM engagement through multiple venues.

The basis for combining robotics and game design in the UGIC study was to provide two types of learning experiences. The researchers wanted to ensure that students had options when they learned to code. In many ways, the drag-and-drop features were easier to use and understand in robotics than they were in SGD as the data showed the learning curve for SGD was high at the outset. Robotics also provided students with opportunities to engage in hands-on activities in addition to computational thinking. If gaming is the only venue for student participation, it is imperative that options are broad enough to ensure that students have some autonomy. Educators can and should learn the value of games in and out of the classroom to engage students in rigorous yet culturally relevant tasks (Leonard, 2019). Finally, the work should be difficult enough to motivate students but not so difficult that they disengage (Renaud & Wagoner, 2011; Webb et al., 2012).

Summary

While the three case studies presented earlier provide a partial glimpse of coding and game design in the extant literature, they also offer some evidence that digital gaming promotes learning in literacy and mathematics in addition to motivation and interest in STEM. Game design can be used to help students find their voice, not only to express creativity, but also their culture, history, and emotions, which can be used to increase cultural awareness and promote equitable STEM education (Leonard & Hill, 2008).

While there are limitations to the Year 2 study, the overall results are positive. Underrepresented student participants (e.g., females, students with disabilities, and students of color) in this study developed robust games that incorporated elements of pop culture and everyday life. Hours spent coding was moderately correlated with mathematics growth. Student growth in mathematics increased significantly when students spent more than 20 hours coding. Rural students' gaming ability improved over time as they learned how to produce different types of games. Some students in a focus group expressed positive attitudes about game design and were motivated to create authentic games. The comments of one female student with disabilities sum it up: *I liked making Frogger and being able to play it. Right now, my favorite game is my first game. It's something like Pac-Man, and I made Snakes. It's really fun!*

References

Barr, V., & Stephenson, C. (2011). Bringing computational thinking to K–12: What is involved and what is the role of the computer science education community? *ACM Inroads*, 2(1), 48–54. https://doi.org/10.1145/1929887.1929905

Chang, K., Wu, L., Weng, S., & Sung, Y. (2012). Embedding game-based problem-solving phase into problem-posing system for mathematics learning. *Computers & Education, 58*(2), 775–786.

Gee, J. P., & Hayes, E. R. (2010). *Women and gaming: The Sims and 21st century learning.* Palgrave McMillan.

Gholson, M. L., & Robinson, D. D. (2019). Restoring mathematics identities of Black learners: A curricular approach. *Theory into Practice, 58*(4), 347–358. https://doi.org/1 0.1080/00405841.2019.1626620

Glaser, B. G. (1965). The constant comparative method of qualitative analysis. *Social Problems, 12*(4), 436–445. http://www.jstor.org/stable/798843

International Society for Technology in Education [ISTE]. (2020). *ISTE standards.* https:// www.iste.org/standards

Jenson, J., Droumeva, M., & Fraser, S. (2016). Exploring medial literacy and computational thinking: A game maker curriculum study. *The Electronic Journal of e-Learning, 14*(2), 111–121.

Kafai, Y. B., & Burke, Q. (2014). *Connected code: Why children need to learn programming.* The MIT Press.

Kitchen, R., & Beck, S. (2016). Educational technology: An equity challenge to the Common Core. *Journal for Research in Mathematics Education, 47*(1), 3–16.

Koutromanos, G., & Avraamidou, L. (2014). The use of mobile games in formal and informal learning environments: A review of the literature. *Educational Media International, 51*(1), 49–65.

Leonard, J. (2009). "Still not saved": The power of mathematics to liberate the oppressed. In D. B. Martin (Ed.), *Mathematics teaching, learning, and liberation in the lives of Black children* (pp. 304–330). Routledge.

Leonard, J. (2019). *Culturally specific pedagogy in the mathematics classroom: Strategies for teachers and students* (2nd ed.). Routledge.

Leonard, J., Barnes-Johnson, J., & Evans, B. R. (2019). Using computer simulations and culturally responsive instruction to broaden urban student's participation in STEM. *Digital Experiences in Mathematics Education, 5*, 101–123. https://doi.org/10.1007/ s40751-018-0048-1

Leonard, J., Buss, A., Gamboa, R., Mitchell, M., Fashola, O. S., Hubert, T., & Almughyirah, S. (2016). Using robotics and game design to enhance children's STEM attitudes and computational thinking skills. *Journal of Science Education and Technology, 28*(6), 860–876. https://doi.org/10.1007/s10956-016-9628-2

Leonard, J., & Hill, M. L. (2008). Using multimedia to engage African-American children in classroom discourse. *Journal of Black Studies, 39*(1), 22–42.

Leonard, J., Mitchell, M., Barnes-Johnson, J., Unertl, A., Outka-Hill, J., Robinson, R., & Hester-Croft, C. (2018). Preparing teachers to engage rural students in computational thinking through robotics, game design, and culturally responsive teaching. *Journal of Teacher Education, 69*(4), 386–407. https://doi.org/10.1177/0022487117732317

Li, Q. (2010). Digital game building: Learning in a participatory culture. *Educational Research, 52*(4), 427–443.

Manches, A., & Plowman, L. (2017). Computing education in children's early years: A call for debate. *British Journal of Educational Technology, 48*(1), 191–201. https://doi. org/10.1111/bjet.12355

Martin, D. B., Gholson, M. L., & Leonard, J. (2010, December). Mathematics as gatekeeper: Power and privilege in the production of knowledge. *Journal of Urban Mathematics Education, 3*(2), 12–24. http://ed-osprey.gsu.edu/ojs/index.php/JUME/article/view/95/57

Newton, K., Leonard, J., Buss, A., Wright, C. & Barnes-Johnson, J. (2020). Learning with robotics and game design in an urban context. *Journal of Research in Technology Education, 52*(2), 129–147. https://doi.org/10.1080/15391523.2020.1713263

Renalli, J. (2008). Learning English with *The Sims:* Exploiting authentic computer simulations games for L2 learning. *Computer Assisted Language Learning, 21*(4), 441–455. https://doi.org/10.1080/09588220802447859

Renaud, C., & Wagoner, B. (2011). The gamification of learning. *Principal Leadership, 12*(1), 57–59.

Repenning, A., Basawapatna, A. R., & Escherle, N. A. (2017). Principles of computational thinking tools. In P. J. Rich & C. B. Hodges (Eds.), *Emerging research, practice, and policy on computational thinking* (pp. 291–305). Springer.

Repenning, A., Webb, D. C., Koh, K. H., Nickerson, H., Miller, S. B., Brand, C., Her Many Horses, I., Basawapatna, A., Gluck, F., Grover, R., Gutierrez, K., & Repenning, N. (2015). Scalable game design: A strategy to bring systematic computer science education to schools through game design and simulation creation. *ACM Transactions of Computer Education, 15*(2), 11.1–11.31.

Ritzhaupt, A. D., Higgins, H., & Allred, B. (2011). Effects of modern educational game play on attitudes toward mathematics, mathematics self-efficacy, and mathematics achievement. *Journal of Interactive Learning Research, 22*(2), 277–297.

Ryoo, J. J. (2019). Pedagogy that supports computer science for all. *ACM Transactions on Computing Education, 19*(4), 36:1–36:23.

Sullivan, F. R., & Heffernan, J. (2016). Robotics construction kits as computational manipulatives for learning in the STEM disciplines. *Journal of Research on Technology in Education, 48*(2), 105–128. https://doi.org/10.1080/15391523.2016.1146563

Unfried, A., Faber, M., Stanhope, D., & Wiebe, E. (2015). The development and validation of a measure of student attitudes toward science, technology, engineering and math (S-STEM). *Journal of Psychoeducational Assessment, 33*(7), 622–639.

Webb, D., Repenning, A., & Koh, K. (2012). Toward an emergent theory of broadening participation in computer science education. In *Proceedings of the ACM Special Interest Group on Computer Science Education Conference* (pp. 173–178). ACM.

Wing, J. M. (2006). Computational thinking. *Communications of the ACM, 49*(3), 33–35. https://doi.org/10.1145/1118178.1118215

Wing, J. M. (2010). *Computational thinking: What and why?* [Unpublished manuscript]. Computer Science Department, Carnegie Mellon University. https://www.cs.cmu.edu/~CompThink/resources/TheLinkWing.pdf

4

USING COMPUTER MODELING AND DRONES TO DEVELOP COMPUTATIONAL THINKING AMONG PREDOMINANTLY BLACK STUDENTS

Jacqueline Leonard

I: *Alright. So, you have a guest speaker today right?*

A: *Yup.*

I: *Cool! I know we kinda talked a little bit about this, but just to recap, what did you learn?*

A: *I learned that about like a billion years ago the earth was basically... nobody was on the earth. I think something came and collided with it and the earth, and they think that that's how the moon was formed, and they know this because when Neil Armstrong and Aldrin? What's his name?*

I: *Buzz.*

A: *Buzz Aldrin went [to the moon]. They took specimen of the rocks, and they examined them, and they were the same as Earth, and that's how they came up with that theory.*

I: *Cool! Here again, I gotta ask, if we had to improve or get somebody else as a guest speaker can you give any suggestions about maybe somebody you would wanna hear from or something we could do to make the experience better?*

A: *I'm not trying to offend him, but I feel like we should get like, umm.... He's great. I'm not complaining, but I feel like we should get like a Black woman because... like for us...I feel like we should get a Black woman because we need somebody... woman-wise, we need like...girls need like somebody to see that somebody like us can be what she's being and could do what she's doing. And for color-wise, because back then Black people didn't have the opportunities that we have now, and I feel like we are still underestimated, and I want us to see... to be inspired and see that umm...that she took her opportunities and made the right choice cause we're usually categorized umm as, like, "we do drugs" and "we don't get out of the hood or streets." And I want us to look at her and be like, "Wow I can do that, and if I work hard and take my opportunities where I see 'em, then I can be where she is or I can be higher."*

The aforementioned student interview occurred on the last day of programming during an afterschool club at an elementary school in Philadelphia. In spring 2019, student participants were recruited from three schools in the same district to participate in the Bessie Coleman Project (BCP), named for the first Black and Native-American female aviator. The student participants were predominantly Black/African American. While the student interview took place during a focus group session in May 2019, it speaks to the importance of Black female role models and how Black lives matter in all facets of life.

Dr. Ron Oliversen, a White male planetary scientist at NASA Goddard Space Flight Center, served as a guest speaker. He was someone Jacqueline (principal investigator) had known for 23 years and had worked with in the past. In fact, Jacqueline was introduced to Dr. Oliversen by Dr. Carol Jo Cranell (November 15, 1938–May 10, 2009), who was a prolific researcher and female role model at NASA Goddard where she began her career as a solar scientist in 1974. Subsequently, a Black female, Dr. Aprille Ericsson-Jackson, began a career as an engineer at NASA Goddard in 1992 (Leonard, 2019). She was the first Black woman to be hired with a doctoral degree in mechanical engineering at this agency. More recently, the U.S. Navy announced its first Black female fighter pilot. After 110 years, Lt. j.g. Madeline Swegle completed Tactical Air (Strike) training in a T-45C Goshawk (Pawlyk, 2020). While drones in the BCP fleet were named after famous Blacks in aviation (e.g., Guion Bluford, Eugene James Bullard, Mae Jemison, Katherine Johnson, Emory Conrad Malick, and others), the importance of Black female speakers should not be glossed over. Black women in STEM can motivate and inspire Black children (Walker, 2014), but more importantly, they possess cultural capital and are living legacies of Black history.

As the young Black girl in the interview pointed out, it is important to change the narrative or "flip the script" on Black students' potential. Too often, the media narrative about Black youth is negative, and too few stories are told about the accomplishments of Black men and women in STEM and other professions (Walker, 2012). The propensity of Black speakers to motivate students to learn STEM in culturally sustaining ways cannot be understated (Bracey, 2013). For example, participating students attending the BCP summer STEM camp in Denver, Colorado, experienced three STEM speakers. While none were female, two speakers were Black. One of the Black male speakers worked as a pilot for United Airlines. He spoke at length about the forces of flight and the need for pilots in the next decade. The third speaker, Capt. Ed Dwight, was the first Black astronaut candidate. He told a fascinating story about his service as an astronaut trainee under President John F. Kennedy and how he later became a world renowned artist and sculptor.

Some of the children's comments about the Black speakers revealed their perspectives on the significance of role models and the salience of race:

> *The sculptor taught us to never give up on [our] dreams and keep going.* (5th grader)

> *Ed Dwight said if you have positive energy, your brain doesn't have to do as much work.* (6th grader)

> *They [Black astronauts] went through tough things, and you have to be a really smart person to get there. Like Black people couldn't fly and now they can.* (7th grader)

In addition to intelligence and positive energy, students' comments emphasized the importance of resilience to accomplish one's goals. Exposing students to STEM professionals and careers was one of the underlying purposes of the BCP. Moreover, we wanted the students to have authentic STEM experiences that would allow them to see themselves as STEM professionals, including pilots, computer scientists, and engineers. We believed these experiences would help the students in the BCP study to develop computational thinking (CT).

Background of the Study

Computational thinking (CT) is a problem-solving process that includes formulating problems, algorithmic thinking, generalization, and learning transfer (Barr & Stephenson, 2011; Repenning, 2012). Using computer programming to develop CT is currently in the forefront of educational reform (Kafai & Burke, 2014; Newton et al., 2020). Some countries are developing content specific courses while others include coding in Information and Communications Technology (ICT) and general technology courses (Falloon, 2016; Moreno-León et al., 2016). As school-based access to technology increases, educational uses of technology are also expanding throughout the curriculum in formal and informal learning environments. With this expansion, there is a critical need to identify the tools, pedagogy, and practices deemed essential for promoting learning and CT among students, especially given the data-rich context in which we currently live.

Computer modeling and drones have been used effectively to develop solutions to complex problems that may be solved with CT. For example, an 11-year-old, African-American boy developed a computer model and patented his idea to prevent infants and toddlers from overheating if left alone in a hot automobile (Fulling, 2017). However, research studies that apply CT to 3D printing and drones are sparse (Bhaduri et al., 2018; Trust & Maloy, 2017) even as these emerging technologies are beginning to become msore prevalent in schools and community-based programs. A review of the literature on innovation technology and engineering applications is presented and subsequently followed by the theoretical framework that undergirds the BCP.

Makerspaces and 3D Printing

Makerspaces are growing in popularity, nationally and internationally, as schools, libraries, museums, and community centers are using large open spaces for children, youth, and adults to engage in do-it-yourself (DIY) projects (Barton & Tan, 2018). A Makerspace is defined as collaborative workspaces where people with shared interests can gather to work on projects that use high tech or no tech tools to share ideas, equipment, and knowledge. Activities, such as 3D printing, can be used in Makerspaces and other settings to break down barriers to learning STEM and to prepare for college and career readiness. However, as Makerspaces and 3D printers become more accessible, attention needs to move beyond making products to incorporating skills that can be acquired during the learning process (Trust & Maloy, 2017). Advocates of 3D printing contend that it fosters innovative experiences that align with constructionism (Papert & Harel, 1991), which is described as learning by doing (diSessa & Cobb, 2004). Three-dimensional printing requires students to engage in computer modeling in Tinkercad or some other program in order to print the artifacts they create. As a result, students engage in spatial reasoning as they manipulate objects in 2D space that will be printed in 3D space. Skills that students learn during this process include self-directed learning, mathematical reasoning, and debugging to resolve printing issues (Trust & Maloy, 2017). Furthermore, students learn 21st-century skills that include "collaboration and teamwork, creativity and imagination, critical thinking, and problem solving" (Trust & Maloy, 2017, p. 256). While disadvantages, such as lack of teacher time, high learning curve, and economic costs, factor into administrative decisions to support 3D printing (Bhaduri et al., 2019; Trust & Maloy, 2017), the disadvantages are outweighed by the benefits, particularly for historically-excluded students who often have limited access to innovative technology during in-school and out-of-school programs (Barton & Tan, 2018; Leonard, 2019).

In a study on 3D modeling and 3D printing, Bhaduri et al. (2019) engaged high school students in a three-week summer academy. The STEM camp took place in a Makerspace that provided access to 3D printers, a high-quality drone, and computers. Using unique storytelling, youth developed a solution to a design challenge to simulate emergency assistance due to a natural disaster. Students used Tinkercad to plan and design an obstacle course for a drone that included creating hook-like attachments and payloads as well as launch pads for take-off and landing. Findings revealed that 3D modeling was motivating for students and strengthened their spatial reasoning and problem-solving skills. These findings concur with the findings in Trust and Maloy's (2017) study, which surveyed K-12 teachers who regularly taught computer-aided design (CAD) programming to 3D print artifacts. Most importantly, the aforementioned studies suggest that modeling artifacts that are personally meaningful to the students and the community (e.g., prosthetic devices, jewelry, reality glasses, and tactile picture books) not only promote ingenuity and creativity but help to revolutionize design and manufacturing in ways that profoundly shift socio-economic and demographic variables by broadening the participation of underrepresented students (Ryoo, 2019; Trust & Maloy, 2017). We build on the aforementioned studies in the BCP by examining students' CT as they engage in the elements of abstraction, automation, and analysis (Repenning et al., 2015).

Using Drones to Engage Students in STEM

Unmanned aerial vehicles (UAVs) or drones have been used by the U.S. military since 2001. Consumers have also used drones, much like kites, for recreational activities. However, as Bhaduri et al. (2019) alluded to in their computer modeling study, drones can be used to access remote or devastated landscapes that are not easily accessible by humans. As a result, aerial photographs and videos of landscapes provide students with opportunities to collect authentic data that can be used to learn science and mathematics (Gillani & Gillani, 2015). In one study that used drones to promote STEM learning and engagement in an after-school club, Gillani and Gillani (2015) provided 10 sixth-grade students with an opportunity to explore the use of drones in their California community. Students learned basic maneuvers to operate drones and then used them to collect data on a nearby lake during a field trip. Aerial video was taken by a factory-programmed drone and then compared with earlier photos that could be found on the Internet to determine the capacity of water at different points in time by observing the difference in sediment on the banks (one acre-foot equals to 325,851 gallons). Students then discussed how much the lake receded over time and how the community could conserve water. Findings revealed UAVs sparked students' interest in science and mathematics. Moreover, students engaged in 21st-century skills through collaboration and teamwork.

In a second study, Bhaduri et al. (2018) engaged middle school students with UAV/drone curriculum during a Saturday school program. The *Engineering Experiences* curriculum was integrated with language arts and mathematics content. The program was offered to two cohorts of middle school students. Cohort 1 consisted of predominantly Latinx children from working-class families in an afterschool program, and Cohort 2 consisted of middle school children from middle-class backgrounds who attended a Saturday academy. The curriculum provided students with three components. The first component consisted of the following: (a) UAV pilot training to learn basic terminology for maneuvers (e.g., pitch, roll, yaw) and performance (i.e., life of battery with and without payload);

(b) a design challenge to retrieve a payload; and (c) aerial video to survey the extent and location of damage in a simulated town called Disasterville. The second component consisted of mission planning to survey the town using recorded video and maintaining flight logs. The final component consisted of completing a storyboard to map out specific air-and-rescue plans for Disasterville.

Results revealed that student interest did not improve significantly but was maintained regardless of cohort. Students' perspectives on engineering attitudes were also more positive at the conclusion of the study—38% positive comments initially compared to 100% positive comments at the end of the study. Qualitative data, however, revealed students' perceptions of drones changed as they realized the utility of using drones as scientific and technological tools for remote sensing. In conclusion, students overwhelmingly wanted to work with drones compared to other possible experiences during the STEM program. However, the researchers learned that students needed ample time to effectively engage in the project and that selecting a "powerful and compelling topic is essential for success" and can serve to drive youth participation (Bhaduri et al., 2018, p. 158). Thus, connecting emerging technologies like UAVs to meaningful activities, such as simulating air-and-rescue missions for natural diseases (e.g., tornadoes, hurricanes, mudslides), holds promise. This study contributed to our thinking about how to use drones purposefully to not only motivate students but to engage them in CT.

Theoretical Framework

The theoretical framework that supports the BCP draws upon two theories—self-efficacy theory (Bandura, 1997) and expectancy-value theory (Wigfield & Eccles, 2000). While self-efficacy measures students' beliefs about success on very specific tasks, expectancy beliefs are measured more broadly to address students' expectations. The value a student puts on a subject or task is predictive of future intentions to participate in similar tasks (e.g., enroll in advanced courses). Expectancy-value theory has a long-standing history in the achievement motivation literature, along with other perspectives on ability beliefs such as self-efficacy theory (Bandura, 1997; Schunk, 2020). Studies on expectancy value conducted with K-12 students reveal findings that are applicable to real-world applications, particularly for students of color and females whose STEM identity and social agency may change as they transition from elementary to middle school. Assessing ability beliefs and values among Black, Latinx, Indigenous students, and females adds to the literature on broadening participation in STEM. The expectancy-value framework has been applied to the subject areas of reading and mathematics. We modified the expectancy-value survey in mathematics to assess students' value of science and technology. Additionally, we examined rural and underrepresented students' self-efficacy in computer programming and CT as it related to the interventions.

The Bessie Coleman Project

The BCP focused on using emerging technology with the goal of enhancing upper elementary and middle school students' computer programming self-efficacy and CT, as well as increasing their motivation to learn STEM in preparation for STEM careers. This study engaged underrepresented students in project-based learning experiences through computer modeling and 3D printing in order to provide coding experiences. Moreover, flight simulation and drones provided students with applications that allowed them to develop new knowledge and understandings of complex systems. However, student participants engaged in culturally relevant interdisciplinary lessons that allowed them to be creators rather than consumers of technology. We were interested in how informal learning environments (i.e., afterschool programs and summer camps) impacted underrepresented students' efficacy, CT, and interest in STEM. However, the primary focus in this chapter is on predominantly Black students.

Research Questions

Based on the goals of the study, we address four research questions in this chapter. The first and second questions, which address student efficacy, CT, and expectancy value, rely on quantitative survey data and classroom observations. The third and fourth questions, which address student attitudes and preferences, rely on qualitative focus group interview data.

1. How did the intervention influence students' self-efficacy in computer programming and CT?
2. How did the intervention influence students' expectancy value of science and technology?
3. How do student participants describe their STEM content learning and attitudes toward STEM after the intervention?
4. What STEM disciplines or careers did student participants identify after the intervention?

Methods

Mixed methods were employed to collect and analyze data in the BCP study. Data were collected by researchers, staff, and teachers. Quantitative measures included a survey instrument that consisted of four subscales (to be described). Qualitative measures consisted of field notes, classroom observations, and focus group interviews (Creswell, 1998). Where possible, ethnography was used to collect qualitative data on the nuances of culture (i.e., Black history, community values, diverse STEM speakers) and place in afterschool clubs and summer camps (Creswell, 1998). Specifically, in this chapter, culture and/or place were used as constructs for learning in each of the urban contexts.

Procedures

In order to implement the BCP study, instructors (classroom teachers and Boys & Girls Club staff) were trained to work with the technological tools. Eighteen hours of face-to-face professional development was offered for classroom teachers at a local school in Philadelphia in January 2019. Additionally, an intensive two-day training session was held for staff at a Boys & Girls Club in Denver in May 2019. Instructors were taught how to operate the 3D printers and fly drones. They were also trained to create 3D objects in Tinkercad and Sculptris and how to use the computer-based flight simulation program. Cultural relevance and place-based education were also addressed in terms of developing lessons where students could print authentic artifacts in 3D and participate in drone challenges that were community based. Students received 30 hours of intervention during afterschool programs or summer camps. Field trips were also planned to provide lesson continuity and opportunities to interact with docents and interactive exhibits to promote career awareness.

Sample

In the tradition of Creswell (1998), we used rich descriptions of schools, learning environments, and participating teachers and students (see Table 4.1). Six teachers and 52 students participated in the study at three schools in Philadelphia (Site 1). Demographics for this cohort were reported as follows: 70% Black; 14% White; 12% Hispanic; 2% Asian; and 2% Native American; 54% male and 46% female. One staff member and a graduate assistant (GA) were the primary instructors at the Boys & Girls Club in Denver. The demographics for this cohort were as follows: 55% Black; 20% two or more races; 15% Hispanic; and 10% White; 65% male and 35% female.

Data Sources and Data Analyses

Quantitative data sources consisted of a STEM survey. The original survey consisted of four constructs: computer programming self-efficacy, computational thinking, science expectancy value, and technology expectancy value. The

TABLE 4.1 Analytic Sample by Study Group and Site

Setting	Site	# Teachers	#Students Enrolled	#Analytic Sample
Philadelphia	1			
	School A	3	33	21
	School B	1	15	9
	School C	2	22	12
Denver	2			
	B & G Club	2	20	16

computer programming self-efficacy (16 items) and computational thinking (5 items) subscales were developed from items found on similar surveys in the literature (Feldhausen et al., 2018; Ketelhut, 2010; Rachmatullah et al., 2020). Responses to the items were based on a 5-point Likert scale (1 = strongly disagree, 2 = disagree, 3 = neutral, 4 = agree, and 5 = strongly agree). Wigfield's and Eccles's (2000) expectancy-value subscale was modified to measure values and beliefs about science and technology. The science expectancy-value scale consisted of 12 items and utilized a 7-point Likert scale (1 = not at all good to 7 = very good). The technology expectancy-value scale was composed of 26 items and utilized a 4-point Likert scale (1 = strongly disagree, 2 = disagree, 3 = agree, and 4 = strongly agree).

Since the survey was comprised of items from multiple published scales, an exploratory factor analysis with varimax rotation was performed on the survey, which resulted in four reliably measured factors: computer programming self-efficacy (4 items), computational thinking (4 items), science expectancy value (4 items), and technology expectancy value (8 items) (see Appendix B). Thus, the survey was revised with only the items that loaded into the four factors. Cronbach's alpha for each of the modified constructs was as follow: computer programming self-efficacy (α = 0.730); computational thinking (α = 0.733); science expectancy value (α = .840); and technology expectancy value (α = 0.822). The abbreviated survey fell within the appropriate range for good reliability. Participants in Year 2 completed the revised survey.

Qualitative data sources included (a) field notes, (b) transcripts of classroom observations that were rated using the Dimensions of Success (DoS) tool (Gitomer, 2014), (c) transcripts of the student interview protocol (see Appendix C), and (d) photographs of student artifacts. Qualitative data allowed the research team to gather information about participating students' STEM engagement and interest, as well as their interactions on field trips and with STEM speakers. However, certain kinds of data collection that were feasible in afterschool programs did not work well at the Boys & Girls Club. For example, the research team was not able to use the DoS tool at the Boys & Girls Club because they helped the staff to facilitate the program.

Quantitative data were analyzed using paired samples *t*-tests to determine whether the interventions led to significant differences in students' computer programming self-efficacy, CT, and expectancy value. In order to analyze qualitative data, transcripts were coded using open coding procedures (Glaser, 1978). The constant comparative method (Strauss & Corbin, 1990) was used after coding the data to find emergent themes and patterns.

Results

The results of the BCP study reported here examined the intervention with two cohorts of underrepresented students in different urban contexts. Student participants at Site 1 in Philadelphia participated in 30 hours of project-based learning during afterschool programs. Students at Site 2 in Denver participated for the same duration during a weeklong summer camp. Because there were slight differences in treatment and the sample sizes differed, the results are reported for the afterschool and summer camp programs separately.

Philadelphia Afterschool Program

Six teachers and 52 predominantly Black students participated in the BCP in Philadelphia (Site 1) in spring 2019. The afterschool program took place in participating teachers' classrooms at three schools two days per week (3:30–5:00 PM). Children in grades five and six used DJI Sparks to learn about UAVs and Flight Simulator X to learn the basics of flying (see Figure 4.1) while third and fourth graders learned how to make objects with Tinkercad and Sculptris (see Figure 4.2).

Self-Efficacy, CT, and Expectancy Value

Students completed pre-post surveys on self-efficacy and expectancy value. The pretest was administered in February 2019 and the posttest in May 2019. The results of the survey constructs were analyzed using a paired *t*-test. As shown in

FIGURE 4.1 Students Working on Flight Stimulator X Program

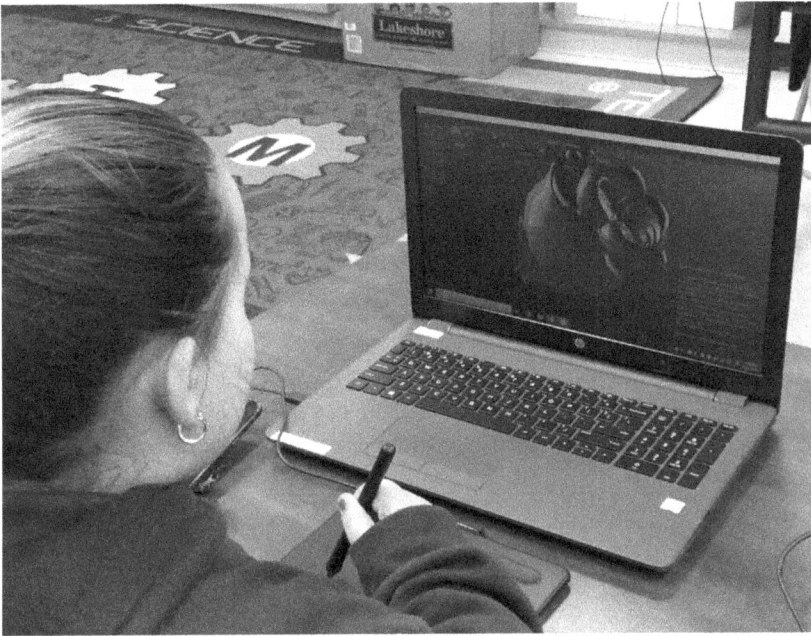

FIGURE 4.2 Student Working on Sculptris

TABLE 4.2 T-test Survey Results: Philadelphia (Site 1)

Site 1	Pretest (Std. Dev.)	Posttest (Std. Dev.)	T-Value	Probability
CP Self-Efficacy (*n*=41)	3.13 (0.86)	3.48 (0.90)	-3.150	0.003★
CT (*n*=41)	3.74 (0.64)	3.63 (0.70)	1.209	0.234
Science Expectancy Value (*n*=41)	5.69 (0.86)	5.36 (1.38)	1.735	0.090
Technology Expectancy Value (*n*=42)	3.56 (0.38)	3.59 (0.36)	-0.691	0.494

★ $p<0.05$

Table 4.2, students' mean efficacy scores on computer programming self-efficacy increased significantly ($p = 0.003$) from pretest 3.13 (std. dev. = 0.86) to posttest 3.48 (std. dev. = 0.90) with a small effect size (Cohen's $d = 0.398$). Survey results also revealed students' CT and science expectancy-value scores declined, but not significantly, after the intervention. Nevertheless, CT scores trended toward a rating of 4 (interpreted as students "agree" that they engage in systematic problem solving). Pre-post science expectancy-value scores ranged from a score of five to six (interpreted as "good" on a 7-point Likert scale). Expectancy-value scores in technology were virtually unchanged from pre- to posttest and trended toward the maximum rating of four (interpreted as "strongly agree" on a 4-point Likert scale).

The results of the survey data were triangulated with qualitative data collected during classroom observations. Two independent raters observed classrooms at each school twice during the spring semester. Analysis of the field notes collected using the DoS observation tool included verbatim transcripts of teacher–student interactions. These data provide additional support to further understand survey results on student engagement in CT. Students were observed using CT during computer modeling (i.e., systematic problem solving) to create artifacts, such as keychains and name tags, using 3D printers. For example, CT is evidenced in a vignette that took place among the teacher [T] and three girls [S_1, S_2, & S_3] at School B.

T: *You will be working on your name tag. Remember you are working with millimeters here.*

S_1: *[Teacher], what are we doing?* [Student came in a little late.]

T: *We are making name tags.* [Leaves to monitor computer simulation in the room next door.]

S_2: *So, are we supposed to just put it [name] on squares?*

S_3: *That's what I did.*

S_2: *If you go in there [other room] and see what the bottom [of the 3D printer] looks like…*

S_3: *So should I make it bigger?*

S_2: *Yes, I would make it a little bigger.*

S_1: [Rotating figures in Tinkercad.]

S_2: *I am going to delete it and redo it.*

S_3: [Also recreated her figure as a heart with her name inside.]

S_2: *Has my name tacked to a nonagon.*

S_3: *Is this good?*

S_2: *If you want to check to see if its good, go and look at the 3D printer.*

S_1: [Draws three small squares on the screen for Tinkercad.]

S_2: *I am going to get a ruler.*

S_1: [Making a pair of eyeglasses.]

S_2: *I just measured. The length is 15(cm) and the width is 23 (cm). So we're making them [name tags] too big. Measures again with a ruler and gets 6 inches by 9 inches for the print bed.* [The three girls work together to convert inches to centimeters.]

Computer modeling allowed this group of students to engage in CT as they created personalized name tags. The girls demonstrated CT as they engaged in abstraction, automation, and analysis (Repenning et al., 2015). First, there was evidence of abstraction when they rotated the figures and made decisions about what shapes to use to make their name tags: *Has my name tacked to a nonagon*. Automation was implicit as one of the girls moved back and forth between the computer and the printer to measure the print bed since 3D printing consists of depositing material layer-by-layer in three dimensions: *If you want to check to see if it's good, go and look at the 3D printer.* Finally, analysis was evident as the girls discussed the size of the name tag, discovered their images were going to be too large, and realized they needed to convert inches to centimeters: *I just measured. The length is 15 (cm) and the width is 23 (cm). So we're making them too big.* Thus, this vignette supports the finding that self-efficacy in computer programming improved significantly and that some students engaged in CT to solve problems that emerged in computer modeling and 3D printing.

Technological Tools and STEM Interest

Four students in Philadelphia at School A (two males and two females) were randomly selected to participate in a focus group interview. After analyzing the qualitative data using open coding procedures, four themes emerged to support the quantitative results: (a) technological tools, (b) field trip, (c) STEM interest, and (d) STEM professionals. Excerpts of students' verbatim responses are reported in Table 5.3. However, every focal student did not have a comment for every theme.

The first theme revealed students enjoyed all of the technological tools. Three focal students mentioned enjoying Tinkercad, and one made a positive comment about Sculptris: *…in Sculptris I liked making it, like rainbow and trying to make a volcano.* Three students mentioned a preference for flying drones. However, one of the younger focal student's comment reflected the desire to fly a drone: *I wish we could learn about the drones, too.* Due to safety precautions, students under the age of 10 did not participate in drone activities.

The second theme related to field trips. Field notes revealed most of the students at all three school sites enjoyed their visit to the Air and Space Museum in Washington, DC. These data were supported by student interviews. Three focal students reported enjoying the hands-on exhibits: *In the Air and Space Museum there were so many cool stuff. You could go check inside the aircraft. We could go touch. We can read. We learned so much.*

For the theme of STEM interest, three focal students had positive attitudes about science and two mentioned liking math. One focal student credited the BCP with learning more about science: *I don't know a lot about science, but this club has helped me to learn a lot about science.…* Another focal student specifically stated that 3D printing influenced his decision to become a scientist: *I like [learning] how to do 3D printing…when I grow up I want to be a scientist or to help people with special needs like, umm, people who have no arms and legs.…* Referring to the vignette given earlier, recall that one of the girls at School B designed a pair of eyeglasses instead of a name tag. This suggests that teachers could tap students' interest in community health and wellness with 3D printing.

The final theme relative to professional speakers also produced interesting results. Two focal students reported that they learned specific science content. One enjoyed learning about light and refraction, and the other enjoyed learning

TABLE 4.3 Philadelphia Focus Group Data

Focus Group Comments	Site 1 (School A)
Technology tools	I liked how we could create things on a computer and that the website [Tinkercad] that we were using, if we were on the same account we could join people and make the same thing and like work together to make something. I wish we could learn about the drones, too.
	Aviation simulator was really fun. Uhm, me and my partner, whichever of my friends was with me that day, whoever wasn't flying tend to be the copilot and like helped turn on the engine and stuff, and helped if we forgot how to do something, and I feel like never mind, it was just really fun. Well in Tinkercad, it was fun to try and figure out how to make a hole because we were making keychains, like how to make a hole in the thing, and in Sculptris I liked making it, like rainbow and trying to make a volcano.
	Tinkercad is super easy to use. You can check all the other people's creations and we can use them, add on to them, take some parts off of them. It's easy to make fun stuff like keychains, and we can use it to raise some money for our school and other stuff.
	Drones are super fun. They are pretty good. They're easy to use. They're one of the best and that's it. [Flight simulation] was super realistic. The [program] had all kinds of planes and helicopters. They were more realistic because you had to fight the draft and you had to learn to land and take off properly.
	I like to learn about planes and stuff, and we got to use aviation stuff, and I was excited to fly the drones.
Field trip	I really loved [the field trip]. In the Air and Space Museum there were so many cool stuff. You could go check inside the aircraft. We could go touch. We can read. We learned so much.
	In the Washington, DC, museum there was like this spaceship that we go on and we could see like the astronauts, but they were like mannequins that we could see and then one of them was floating and that was really fun.
	[The field trip] was really fun because uhm, we got to like learn about stuff in space, and in the planet's little exhibit, I was kinda mad because they took off Pluto. It's a dwarf planet.
STEM interest	Because I like science. It's my favorite subject in school and umm and because I like to learn more about umm thing I haven't learned before like… and I wanted to know more about it and how it worked and everything.
	My favorite thing to do at school is math. Science is interesting to me. I don't know a lot about science, but this club has helped me to learn a lot about science and to learn about the museums that we went to so it pretty fun.

(Continued)

TABLE 4.3 (Continued)

Focus Group Comments	Site 1 (School A)
	Because I like science and math and I like [learning] how to do 3D printing…when I grow up I want to be a scientist or to help people with special needs like, umm, people who have no arms and legs so we could 3D print some arms and legs so we could help them with their lives and other stuff.
STEM professionals or STEM careers	Oh, I like the things the scientist handed out…because it like has rainbow things so whenever we look at the [light] through the screen it had rainbow lines in it so that was really fun to look at.
	So we had a [speaker] from [NASA], and he talked about the moon and the astronauts that went there and the solar eclipse and why we have months and why we have years, etc.
	They [Black astronauts] went through tough things, and you have to be a really smart person to get there. Like Black people couldn't fly and now they can.

about the moon launch and solar eclipses: …*he [planetary scientist] talked about the moon and the astronauts that went there and the solar eclipse and why we have months and why we have years.*

Denver Summer STEM Camp

The summer camp took place at a Boys & Girls Club in Denver in June 2019. Children were recruited from the nearby community, which was predominantly Black and Latinx. Some of the student participants were siblings or cousins while others were friends that lived in the same neighborhood. Instruction took place in a multiple purpose room that had plenty of space for laptop computers and two 3D printers. The program ran from 9 AM to 3 PM each day. Tinkercad was used to engage in computer modeling and 3D printing. Students created keychains and dinosaurs as artifacts. Dinosaurs roamed in the Denver basin during the Cretaceous period; therefore, lessons that allowed students to 3D print dinosaurs were placedbased (see Figure 4.3).

To learn about aviation, students used Flight Simulator X and flew DJI Mavic Pros at a nearby park. They participated in four drone lessons where they learned how to (a) take off and land the drone on a target; (b) travel around an obstacle course; (c) use the indicators on the drone to measure the distance from the ground; and (d) use stopwatches to time different flight paths.

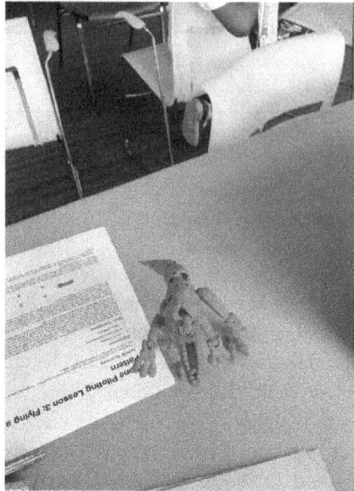

FIGURE 4.3 3D Image of T-Rex

Self-Efficacy, CT, and Expectancy Value

Students in the Boys & Girls Club completed the same pre-post survey as students in Philadelphia. However, they took the pretest on Monday and the posttest on Friday of the same week, which resulted in a shorter timeframe for post-testing. A paired *t*-test was used to analyze the survey constructs (see Table 4.4). The results show that computer programming self-efficacy was statistically significant ($p = 0.046$) with a small effect size (Cohen's $d = 0.434$) as mean scores increased from 3.04 (Std. Dev. = 0.97) on the pretest to 3.49 (Std. Dev. = 0.92) on the posttest. Moreover, CT was statistically significant ($p = 0.008$) with a medium effect size (Cohen's $d = 0.566$) as mean scores increased from 3.32 (Std. Dev. = 0.67) on the pretest to 3.68 (Std. Dev. = 0.60) on the posttest. Pre-post scores on science and technology expectancy value increased but not significantly. Yet, science expectancy value trended toward a score of five (interpreted as feeling "good" on a 7-point Likert scale), and the technology expectancy-value scores trended toward a score of three (interpreted as "agree" on a 4-point Likert scale).

TABLE 4.4 T-test Survey Results: Denver (Site 2)

Site 2	Pretest (Std. Dev.)	Posttest (Std. Dev.)	T-Value	Probability
CP Self-Efficacy (*n*=16)	3.04 (0.97)	3.45 (0.92)	-2.176	0.046★
CT (*n*=17)	3.32 (0.67)	3.68 (0.60)	-3.058	0.008★
Science Expectancy Value (*n*=16)	4.44 (1.27)	4.81 (1.16)	-1.369	0.191
Technology Expectancy Value (*n*=16)	3.15 (0.62)	3.26 (0.54)	-0.739	0.471

Despite the small sample size and different learning contexts at the Boys & Girls Club, the results in Denver were very encouraging.

Technological Tools and STEM Interest

Four students of equal gender were randomly selected to participate in a focus group interview. Data were transcribed and coded for patterns and analyzed for themes. The themes that emerged in Philadelphia also emerged in Denver with the exception of one additional theme—starting a company. Recall that not every student made comments that were indicative of each theme. Moreover, focal students could make multiple comments about a particular theme.

The first theme revealed that two focal students enjoyed drones and one focal student enjoyed computer modeling: *Computer modeling was really fun and interesting but really hard.* However, the unique theme of starting a business was related to technological tools. Students realized that computer modeling was purposeful as artifacts that represented a particular company could be 3D printed. Three of the four focal students made comments about design or CAD and two referenced 3D printing: *We thought of a company, modeled something we wanted to sculpt, then we did Tinkercad and reimagined our thing then printed our idea.*

Relative to the second theme, Denver students enjoyed their field trip to the Wings Over the Rockies Air and Space Museum. Two focal students reported that they enjoyed sitting in a F-14 Tomcat: *I like how we got to sit in a jet or plane.* Two were enthralled with the museum's flight simulator: *I liked how we could do the simulator in space. Just like there but not there with the flight simulator.*

In terms of STEM interest, three of the four focal students were interested in technology, and one noted the importance of gender equity in this field: *I think it's important for women to know about technology.* One of these focal students also mentioned an interest in science, while another specifically mentioned wanting to become a computer scientist.

Finally, the timing and flexibility of the summer camp allowed for more speakers. While all three speakers were male, evidence suggests that students learned from each of them. Specifically, students reported learning science content that was related to the refraction of light and aerodynamics. However, the most salient part of this theme was the resilience and brilliance of Black STEM professionals: *Like Black people couldn't fly and now they can. I've seen the building [Ed Dwight] built by City Park. I don't know how he did all that stuff!*

The focus group data also provided additional insight for understanding participating students' CT. Elements of Repenning et al.'s (2015) CT model were visible during computer modeling activities: *We modeled something we wanted to sculpt* (abstraction); *…then we did Tinkercad and reimagined our thing then printed our idea* (abstraction and automation); *Say you're trying to make a circle but only have a square. You can use a hole around the square by pulling it….* (analysis) (Table 4.5).

TABLE 4.5 Denver Focus Group Data

Focus Group Comments	Site 2 (B & G Club)
Technology tools	I liked how we got learn about the airplanes, spaceships and fly the drones. At home, my uncle already have a drone, and I learned from him.
	I liked flying the drones and now I can fly mine. I didn't know how to get it off the ground. Like right now, my dad came out and I crashed it. I had to get going and I got mad. But, I hope it still works.
	Computer modeling was really fun and interesting but really hard. You have to know what to do and how to do it. Say you're trying to make a circle but only have a square. You can use a hole around the square by pulling it.
Starting a company	We thought of a company, modeled something we wanted to sculpt, then we did Tinkercad and reimagined our thing then printed our idea.
	Today, [we] did our favorite company and what we think. We get to 3D print our company, that we know and draw about it and write about it.
	Basically, we got to choose a company, then the company we chose we wrote what it does. I chose Nintendo, and wrote games, controller. We were able to design something…we wanted for the company.
Field trip	I like how we got to sit in a jet or plane. It was really tight, but that's a good thing….
	We got to go into the cockpit….
	We got to play the simulator and I got to do it once.
	I liked how we could do the simulator in space. Just like there but not there. With the flight simulator.
STEM interest	I came to the camp to learn some stuff about technology and spacecraft.
	I think it's important for women to know about technology.
	I came to camp to get better with science and technology….
	I want…to become a computer scientist.
STEM professionals	I didn't know there were different kinds of lights to make different colors of the rainbow. Captain D. [told us] all about airplanes and rocket ships. Also about air, and how it goes over our hands.
	I learned about sculptures, and I've seen the building [Ed Dwight] built by City Park. I don't know how he did all that stuff!
	I really like how Captain D. talked about flying jets and what he does. Ed Dwight talked about his sculpting. The first guy talked about lights and different perspectives.

Discussion

The results of the BCP study in urban contexts revealed four findings. The first finding relates to computer programming and CT. Computer programming self-efficacy scores increased significantly from pre-posttest among students in both cohorts. However, CT scores declined slightly among students in Philadelphia but increased significantly among students in Denver. We attribute the significance in computer programming to high interest in computer modeling with Tinkercad and Sculptris, as well as Flight Simulator X (Trust & Maloy, 2017), and the significance in CT to implementing the full intervention in Denver. Recall that Philadelphia students participated in computer modeling/3D printing or flight simulation/drones. Yet, students' pretest scores on CT were higher in Philadelphia than Denver at the outset and may have suffered from ceiling effects.

While qualitative data from focus groups support this finding, we also learned that it is important for students to create meaningful products in Makerspace settings (Barton & Tan, 2018; Bhaduri et al., 2019). Excerpts from one classroom observation and focus groups revealed that some students were interested in creating products that consumers needed. One student designed a pair of eyeglasses and another student wanted to become a scientist or mechanical engineer to create prosthetics. If products such as eyeglasses, hearing aids, and prosthetics can be 3D printed, these aids will become more accessible and affordable in underserved communities. Starting a company and developing 3D products could also promote entrepreneurship in Black and Latinx communities. This connection to entrepreneurship was a real-world activity that went beyond making keychains (Barton & Tan, 2018) and yielded important implications, such as community health and wellness, as well as thriving businesses, which are pivotal to the Black Lives Matter movement and other social justice initiatives (Taylor, 2016). Furthermore, teachers can use the ideas that emerged from students in this study to engage them in authentic culturally relevant and/or place-based activities. Like the student who patented a device to prevent infant and toddler deaths in cars (Fulling, 2017), evidence suggests that some students may want to make products that matter to their lives and communities.

The second finding relates to students' expectancy value in science and technology. Results of paired t-tests revealed that students' science expectancy scores decreased at Site 1 (Philadelphia) while they increased at Site 2 (Denver). While these results were not statistically significant, one explanation is that additional STEM speakers may have influenced Denver participants' expectancy value and interest in science. Recall that Philadelphia participants had only one STEM speaker and Denver participants had three. Another explanation is that students' pretest scores on science expectancy value were higher in Philadelphia than Denver at the outset and may have been influenced by ceiling effects.

The third finding is that students enjoyed using the technological tools associated with computer modeling and 3D printing as well as drones and flight simulation. While some focal students commented on the difficulty associated with computer modeling and drones, they were not deterred. Two students mentioned having experiences with their personal drone under the supervision of a relative. Like computer modeling, however, drone usage should be meaningful (Bhaduri et al., 2018). Once students learn the basics of flying, they should learn how to use UAVs to help their communities (Gillani & Gillani, 2015). For the final year of the study, lessons are being developed to use the drones to find suitable space in urban cities for economic development and to collect aerial video in rural settings to study climate change.

The fourth finding is that role models matter. Students learned a great deal from each of the STEM speakers. Comments from focus group students ranged from learning what was possible for Black people to statements of resilience. Nevertheless, one Black girl's request for a Black woman to serve as a STEM speaker is not without merit. Several women scholars have shared unique stories of racism and sexism in institutions of higher education (Johnson et al., 2019; Leonard, 2012; McGee, 2016). Therefore, it is imperative that Black girls hear stories of resilience and persistence from women in STEM who have not only endured racism but succeeded in spite of it. In the final year of the study, underrepresented female STEM professionals will be recruited to serve as guest speakers.

Summary

While the BCP was proposed as a three-year study (2018–2021), only two years of data have been collected and analyzed so far. Due to COVID-19, the project was paused in Year 3. The project team developed online modules that students could do from home (see Appendix D for Artifacts in Space). These modules will allow the research team to expand the project to more students beyond the initial research sites. Field-testing has shown excitement and student interest in these modules.

Research that examines underrepresented students' self-efficacy in computer programming, CT, and expectancy value in STEM is sparse in informal learning environments. The data reported here reveal that student engagement in emerging technologies in out-of-school programs holds promise for broadening underrepresented students' interest in STEM. As the BCP study demonstrates, Black children and youth are capable STEM learners who are interested in solving real-world problems that face their communities (Fulling, 2017). For Black students who have been underestimated and marginalized in STEM, "making" can be an empowering experience that leads to community uplift as well as future success in STEM (Barton & Tan, 2018; Leonard et al., 2017).

References

Bandura, A. (1997). *Self-efficacy: The exercise of control.* W. H. Freeman.

Barr, V., & Stephenson, C. (2011). Bringing computational thinking to K-12: What is involved and what is the role of the computer science education community? *ACM Inroads, 2*(1), 48–54. doi:10.1145/1929887.1929905

Barton, A. C., & Tan, E. (2018). *STEM-Rich maker learning.* Teachers College Press.

Bhaduri, S., Gendreau, A., Koushik, V. S., Sumner, T., Ristvey, J., & Russell, R. (2018). Promoting middle school students' motivation and persistence in an after-school engineering program. In J. Barnes-Johnson & J. Johnson (Eds.), *STEM 21: Equity in teaching and learning to meet global challenges of standards, engagement and transformation* (pp. 138–162). Peter Lang.

Bhaduri, S., Sumner, T., & Van Horne, K. (2019, May 4–9). *Designing an informal learning curriculum to develop 3D modeling knowledge and improve spatial thinking skills* [Paper presentation]. CHI Conference on Human Factors in Computing Systems Extended Abstracts, Glasgow, Scotland UK. https://doi.org/10.1145/3290607.3299039

Bracey, J. (2013). The culture of learning environments: Black student engagement and cognition in math. In J. Leonard & D. B. Martin (Eds.), *The brilliance of black children in mathematics: Beyond the numbers and toward new discourse* (pp. 171–194). Information Age Publishing.

Creswell, J. W. (1998). *Qualitative inquiry and research design choosing among five traditions.* Sage Publications.

diSessa, A., & Cobb, P. (2004). Ontological innovation and the role of theory in design experiments. *Journal of the Learning Sciences, 17*(4), 465–516.

Falloon, G. (2016). An analysis of young students' thinking when completing basic coding tasks using Scratch Jnr. on the iPad. *Journal of Computer Assisted Learning, 32*(6), 576–593. doi:10.1111/jcal.12155

Feldhausen, R., Weese, J. L., & Bean, N. (2018). Increasing student self-efficacy in computational thinking via STEM outreach programs. In *Proceedings of the 49th ACM Technical Symposium on Computer Science Education* (pp. 302–207). https://doi.org/10.1145/3159450.3159593

Fulling, J. (2017, July 19). Kid's device aims to end hot car deaths. *USA Today,* p. 6B.

Gillani, B., & Gillani, R. (2015). From drought to drones: An after-school club uses drones to learn about environment science. *Science & Children, 53*(2), 50–54.

Gitomer, D. (2014). *Development of the Dimensions of Success (DoS) observation tool for the out of school time STEM field: Refinement, field-testing and establishment of psychometric properties.* Program in Education, Afterschool, and Resilience Institute. http://www.pearweb.org/tools/dos.html

Glaser, B. G. (Ed.). (1978). *Theoretical sensitivity: Advances in the methodology of grounded theory* (Vol. 2). Sociology Press.

Johnson, H. J., Dunleavy, T. K., & Joseph, N. M. (2019). Noyce at Vangerbilt: Exploring the factors that shape the recruitment and retention of Black teachers. In J. Leonard, A.

C. Burrows, & R. Kitchen (Eds.), *Recruiting, preparing, and retaining STEM teachers for a global generation* (pp. 58–77). Brill/Sense.

Kafai, Y. B., & Burke, Q. (2014). *Connected code: Why children need to learn programming.* The MIT Press.

Ketelhut, D. J. (2010). Assessing gaming, computer and scientific inquiry self-efficacy in a virtual environment. In L. Annetta & S. C. Bronack (Eds.), *Serious educational game assessment: Practical methods and models for education games, simulations, and virtual worlds* (pp. 1–18). Sense Publishers.

Leonard, J. (2012). Er'body talkin' 'bout social justice ain't goin' there. *Journal of Urban Mathematics Education, 5*(2), 18–27.

Leonard, J. (2019). *Culturally specific pedagogy in the mathematics classroom: Strategies for teachers and students* (2nd ed.). Routledge.

Leonard, J., Walker, E. N., Cloud, V. R., & Joseph, N. M. (2017). Mathematics literacy, identity resilience, and opportunity sixty years since *Brown v. Board*: Counternarratives of a five-generation family. In J. Ballenger, B. Polnick, & B. Irby (Eds.), *Women of color in STEM: Navigating the workforce* (pp. 79–107). Information Age Publishers.

McGee, E. O. (2016). Devalued Black and Latino racial identities: A by-product of STEM college culture? *American Educational Research Journal, 53*(6), 1626–1662. https://doi.org/10.3102/0002831216676572

Moreno-León, J., Robles, G., & Román-González, M. (2016). Code to learn: Where does it belong in the K-12 curriculum? *Journal of Information Technology Education: Research, 15*, 283–303. http://www.informingscience.org/Publications/3521.

Newton, K., Leonard, J., Buss, A., Wright, C., & Barnes-Johnson, J. (2020). Learning with robotics and game design in an urban context. *Journal of Research in Technology Education, 52*(2), 129–147. doi:10.1080/15391523.2020.1713263

Papert, S., & Harel, I. (1991). Situating constructionism. In I. Harel & S. Papert (Eds.), *Constructionism: Research reports and essays, 1985-1990* (n.p.). Ablex Publishing.

Pawlyk, O. (2020). The U.S. Navy has its first Black female tactical jet pilot. *Military.com*, n.p. https://www.military.com/daily-news/2020/07/10/after-110-years-of-aviation-navy-get-its-first-black-female-fighter-pilot.html

Rachmatullah, A., Wiebe, E., Boulden, D., Mott, B., Boyer, K., & Lester, J. (2020). Development and validation of the Computer Science Attitudes Scale for middle school students (MG-CS attitudes). *Computers in Human Behavior Reports 2*. https://doi.org/10.1016/j.chbr.2020.100018

Repenning, A. (2012). Education programming goes back to school: Broadening participation by integrating game design into middle school curricula. *Communication of the ACM, 55*(5), 38–40.

Repenning, A., Webb, D. C., Koh, K. H., Nickerson, H., Miller, S. B., Brand, C., Her Many Horses, I., Basawapatna, A., Gluck, F., Grover, R., Gutierrez, K., & Repenning, N. (2015). Scalable game design: A strategy to bring systematic computer science education to schools through game design and simulation creation. *ACM Transactions of Computer Education, 15*(2), 11.1–11.31.

Ryoo, J. J. (2019). Pedagogy that supports computer science for all. *ACM Transactions on Computing Education, 19*(4), 36:1–36:23.

Schunk, D. H. (2020). *Learning theories: An educational perspective* (8th ed.). Pearson.

Strauss, A. L., & Corbin, J. (1990). *Basics of qualitative research: Grounded theory procedures and techniques.* Sage.

Taylor, K-Y. (2016). *From #BlackLivesMatter to Black liberation.* Haymarket Books. https://doi.org/10.1080/07380569.2017.1384684

Trust, T., & Maloy, W. R. (2017). Why 3D print? The 21st-Century Skills students develop while engaging in 3D printing projects. *Computers in the Schools, 34*(3), 253–266.

Walker, E. N. (2012). *Building mathematics learning communities: Improving outcomes in urban high schools.* Teachers College Press.

Walker, E. N. (2014). *Beyond Banneker: Black mathematicians and the paths to excellence.* State University of New York Press.

Wigfield, A., & Eccles, J. S. (2000). Expectancy-value theory of achievement motivation. *Contemporary Educational Psychology, 25*(1), 68–81. doi:10.1006/ceps.1999.1015

5

FACILITATING COMPUTATIONAL PARTICIPATION, PLACE-BASED EDUCATION, AND CULTURALLY SPECIFIC PEDAGOGY WITH INDIGENOUS STUDENTS

Jacqueline Leonard

Jacqueline, first author of this text, was extremely excited about her trip to Halifax, Nova Scotia, as an extension of her work as the Fulbright Canada Research Chair in STEM Education at the University of Calgary. Landing in Halifax in early December 2018, Jacqueline rented an SUV and traveled 2 hours by car to Antigonish, where she met with faculty and K-12 mathematics teachers at St. Francis Xavier University. The next day, she traveled with two researchers from the university—one First Nations and one White female—to Cape Breton to observe Mi'kmaq students at an elementary school. Upon arrival, the team spent about 40 minutes with the school administrator. Another 20 minutes was spent touring the school. During the tour, Jacqueline noticed a large photograph of a Native woman and Queen Elizabeth II, who was holding a basket. The woman in the photograph was a survivor of the Indian Residential Schools. The basket was given to the queen as a gesture of goodwill after Prime Minister Stephen Harper offered First Nations peoples a formal apology in 2008 for forcing their children to attend residential schools for the purpose of assimilation. Ten percent of the children housed at the residential school in Shubenacadie, Nova Scotia, from 1930 to 1967, were Mi'kmaq, and the names of survivors were listed on plaques at the elementary school in Cape Breton. After the tour, Jacqueline observed Mi'kmaq students in a fifth-grade mathematics class taught by the St. Francis Xavier University professor/researcher. Her teaching practices included use of Mi'kmaq ways of knowing and verbing (to be described). As a result of this school visit, Jacqueline learned decolonized research methods that guided her work with Arapaho children and youth on the Wind River Indian Reservation in Wyoming.

Inclusion and STEM

The computer science and information technology field is expected to grow 16% in this decade (Bureau of Labor Statistics [BLS], U.S. Department of Labor, 2016). Yet, less than 11% of African Americans (5.41%), Hispanics (5.20%), and Native Americans (0.07%) combined were represented in the field of computer science in 2015 (National Science Foundation [NSF], 2017). Efforts to broaden participation in science, technology, engineering, and mathematics (STEM) among historically underrepresented students continue to be dismal and have not led to increased participation in computer and information science careers. Statements such as, *"Well, we don't have any Native Americans at our company"* will continue to abound as long as diversity and inclusion policies have no teeth (Florentine, 2019, p. 3). While colonialism and overt racism like that described in the vignette may no longer occur, implicit bias and systemic racism continue to hinder opportunities for Indigenous and other underrepresented minorities to fully participate in STEM.

To address these problems, President Obama launched the CS4All Initiative in 2016. It was a bold initiative that made preparing underrepresented students with computer science (CS) and information and communications technology (ICT) skills a national priority. Given the stark statistics mentioned earlier, building STEM pathways for Indigenous students (Alaska Natives, Native Americans, and Pacific Islanders) is a daunting task. Barriers to participation in STEM in general and computer science in particular include limited access to computers and the Internet, economic factors, geographic isolation, and cultural incongruity between Western and Indigenous ways of knowing (Brayboy & Maughan, 2009; Minero, 2019; Tzou et al., 2019). Yet, earning a degree in computer science offers Indigenous peoples opportunities to work remotely, preserve their culture, and make an impact in their communities and the world at large (Florentine, 2019; Minero, 2019).

Corporations like Intel have made some progress toward increasing Native-American representation in computer science. Intel's Native Coders program provided hundreds of Native-American high school students with culturally responsive curriculum to achieve the goal of full representation in a diverse workforce by 2020 (Florentine, 2019). The Native Coders program, which began in 2015, provided computer science educators and a computer lab at three Navajo high schools in Arizona. Partnering with the American Indian Science and Education Society (AISES), Intel convened tribal leaders, non-profits, caregivers, and students together to build strategies, create products, and promote programs that leveraged STEM with Indigenous traditions (Florentine, 2019). Thus, the Native Coders program provides a counternarrative that has the potential to change negative perceptions about Indigenous peoples' place in ICT.

The purpose of this chapter is to provide an additional counternarrative about Indigenous students, who participated in the Bessie Coleman Project (BCP). Indigenous children and youth learned about the principles of flight, participated in flight simulation, and used drones to enhance their knowledge of water, land, and plants on the Wind River Indian Reservation (WRIR). The privilege to conduct research on the reservation was the result of the research team's efforts to build relationships with faculty, school administrators, and teachers from Indigenous backgrounds since 2012. Prior work with Indigenous teachers and students, who had participated in an earlier study on robotics and game design (see Leonard et al., 2016, 2018), helped to provide access to Indigenous teachers and staff to conduct the BCP. The goal of the BCP was to expose Indigenous children and youth (and other underrepresented students) to STEM and STEM careers using unmanned aerial vehicles (UAVs), commonly called drones, and flight simulation as a springboard to deepen STEM content knowledge and just have fun.

Positioning Indigenous Students as Able Computer Science Learners

According to Varma (2009), approximately 2.5 million Native Americans live in the U.S. Twenty percent of Native Americans live on more than 300 reservations (Varma, 2009) in roughly 573 tribal communities (Florentine, 2019). Historically, the language, beliefs, culture, and practices of Indigenous peoples have been viewed through a deficit lens. The Indian Boarding/Residential Schools, described in the opening vignette, were used to "kill the Indian and save the man" Brayboy (2005, p. 430). Specifically, Indian Boarding/Residential Schools were used to deprive Indigenous children of their language and culture and compel the assimilation of Western ways of speaking and knowing (Hipp, 2019).

In STEM education, understanding the importance of other ways of knowing is crucial to helping students to cross cultural borders associated with learning STEM (Aikenhead, 2001; Leonard, 2019). Theories that draw upon the values of Indigenous culture, which are based on family and community, and preserve cultural identity are needed to broaden the participation of Indigenous peoples in STEM (Tzou et al., 2019). Rather than adhering to deficit theory, researchers and teachers should view Native-American students' cultures as assets upon which to build new knowledge (Bracey, 2013; Leonard, 2019; Leonard et al., 2018).

Indigenous Knowledge Systems

Brayboy and Maughan (2009), Bang et al. (2012), and other Indigenous scholars have enlightened mainstream educational researchers about Indigenous knowledge systems (IKS). Simply put, IKS are "rooted in the lived experiences of peoples" where the "experiences highlight the philosophies, beliefs, values, and educational processes of entire communities" (Brayboy & Maughan, 2009, p. 3). Using sociocultural perspectives on learning and development as the underlying framework, IKS consist of epistemologies (what is *knowledge*), ontologies (what is *real*), axiologies (what is *valued*), and pedagogies (what is *taught*) (Brayboy & Maughan, 2009; Tzou et al., 2019). Grounding STEM learning in Indigenous oral traditions can create powerful stories that complement Indigenous and Western knowledge systems in generative ways (Brayboy & Maughan, 2009; Tzou et al., 2019).

Two studies that complement IKS in STEM education offer not only excellent STEM examples but also additional terminology. In the first study, Borden (2011) introduced the term *verbification* in mathematics to support student learning among Mi'kmaq students in Nova Scotia. *Verbification* became a way of talking about mathematics in Mi'kmaq classrooms in a non-objective manner. For example, pyramids "form a triangle" or "come to a point" while cubes "sit still." In other words, mathematics is an active (e.g., in motion) rather than a static process. The Mi'kmaq identify as Aboriginal peoples whose approach to education consists of anti-oppressive worldviews that include IKS (Borden, 2011). Borden offered a nuanced research approach that included respect for culture, language, ways of knowing, and building relationships among insiders and outsiders called *fictive kin* (Lipka et al., 1998). Borden's culturally based research approach consists of three principles: resistance, political integrity, and privileging Aboriginal voices through moral praxis and relational context (Borden, 2011). The method used to teach mathematics in the Mi'kmaq community was termed *mawikinutimatimk*, which means "coming together to learn together" (Borden, 2011, p. 9).

Similarly, a second study on STEM-Art relied on intergenerational work in a Makerspace that involved using materials and robotics in ways that allowed Indigenous families to engage in what Tzou et al. (2019) called *survivance* and *presencing*. Indigenous *survivance* (survival and resilience) is the active *presencing* of IKS through the use of stories, while instantaneously rejecting the erasure of Indigenous culture and values. Moreover, the study participants engaged in *family-making* that included forging Indigenous identities that were based on "relationships with lands, waters, and ways of knowing and being" (Tzou et al., 2019, p. 309). Case studies of two families revealed how they intertwined robotics and STEM-Art with stories to transform the *making experiences* for Indigenous purposes rather than learning technology in a vacuum. In other words, the families engaged in computational participation that expanded knowledge of Indigenous "nature-cultural relations in a way that [was] consequential and liberatory for family and community ends" (Tzou et al., 2019, pp. 323–324). Computational participation involves understanding human behavior as a community rather than an individual perspective to solve common problems and develop solutions in a systematic way (Kafai & Burke, 2014). The aforementioned studies enlarged our understanding of computational thinking to encompass computational participation to guide our work with Indigenous children and youth in Wyoming.

Theoretical Framework and Guiding Principles

The theoretical framework that supports IKS is Tribal Critical Race Theory (TribalCrit). TribalCrit is a branch of Critical Race Theory (CRT) that provides a framework to support Indigenous values, culture, and identity (Brayboy, 2005). CRT began as a form of legal scholarship that examined how the law intersected with race to show the limitations of policies that were based on color blindness and meritocracy (Anderson, 2019; Bell, 1987; Crenshaw, 1988). There are nine TribalCrit tenets. Four tenets[1], which are common to CRT, situate TribalCrit as a form of resistance by naming the root causes of racism: (a) colonization is endemic to society; (b) U.S. policies toward Indigenous people are rooted in imperialism, White supremacy, and a desire for material gain; (c) Indigenous peoples occupy a liminal space that accounts for both the political and racialized natures of [their] identities; and (d) government policies and educational policies toward Indigenous peoples are intimately linked around the problematic goal of assimilation (Brayboy, 2005, p. 429). Therefore, protecting and preserving the sovereignty and autonomy of Indigenous peoples are paramount to successfully implementing anti-oppressive and social justice-oriented STEM programs.

Brayboy (2005) also delineates four tenets that describe what should be learned or attended to in TribalCrit: (e) Indigenous peoples have a desire to obtain and forge tribal sovereignty, tribal autonomy, self-determination, and self-identification; (f) the concepts of culture, knowledge, and power take on new meaning when examined through an Indigenous lens; (g) tribal philosophies, beliefs, customs, traditions, and visions for the future are central to understanding the lived realities of Indigenous peoples, but they also illustrate the difference and adaptability among individuals and groups; and (h) stories are not separate from theory; they make up theory and are, therefore, real and legitimate sources of data and ways of being (pp. 429–430). Thus, STEM activities must be meaningful, purposeful, and counterhegemonic to support Indigenous culture, beliefs, customs, and/or traditions. In other words, Indigenous children and youth should see themselves in the curriculum and understand how it is relevant and makes a difference for the community at large.

The final tenet provides insight into how theory and practice should be viewed: (i) theory and practice are connected in deep and explicit ways such that scholars must work toward social change (p. 430). One way to conduct this work is through counternarratives. CRT, including TribalCrit, legitimizes the experiential knowledge of people of color and supports counterstories that challenge racial stereotypes (Solórzano & Yosso, 2002; Yosso, 2006). Consequently, counternarratives that support and enhance Indigenous students' learning and success in STEM should lead to systematic change that broadens the participation of

Indigenous peoples in STEM (Bang & Marin, 2015). Yet, hegemonic and systemic barriers to education, as well as economic gaps and cultural dissonance, have limited opportunities for Native Americans and other underrepresented students in STEM and, in many cases, their voices have been marginalized (Brayboy, 2005; Stavrou & Miller, 2017). Thus, there is a need for research projects like the BCP to counter hegemonic practices with social justice.

A brief history of the WRIR is presented to recognize some of the egregious acts, perpetuated by the U.S. government, that Indigenous peoples in Wyoming (and elsewhere) have endured. This history is followed by two of the guiding principles that ground our work with Indigenous students—place-based education (PBE) and culturally specific pedagogy (CSP).

History of the Wind River Indian Reservation

The WRIR (pop. 26,500) is located near the town of Riverton (pop. 11,000) in central-western Wyoming near the eastern slope of the Wind River Mountains. The WRIR was established in 1868 under the Second Treaty of Fort Bridger between the Eastern Shoshone and the U.S. government (Flynn, 2008). In 1874, Congress decreed treaties with Indian tribes illegal. However, due to the discovery of gold, the U.S. government negotiated with the Shoshone to cede land back to the government (i.e., Brunot Cession) in exchange for livestock, school funding, and cash (Flynn, 2008). Throughout the years, additional land was sold to White settlers under allotment policies, such as the General Allotment Act of 1887 (Flynn, 2008). At the present time, the WRIR is approximately 2.3 million acres or about 3,800 square miles (Flynn, 2008).

In 1878, the Arapaho tribe was brought to the WRIR without the Shoshone's consent. Through the Fort Laramie Treaty of 1851, the Arapaho were promised land in northern Colorado and western Kansas and Nebraska. However, the treaty was not honored, and the Arapaho never received these lands. Instead, nearly 1,000 Arapahos were taken to the WRIR under military escort as a temporary residence (Flynn, 2008). In 1927, the Shoshone successfully sued the U.S. government for giving the Arapaho a portion of their lands without permission or compensation (Flynn, 2008). Consequently, the Arapaho, who outnumber the Shoshone two-to-one, became legitimate residents of the reservation. Thus, the WRIR is home to the Eastern Shoshone and Northern Arapaho, which is the only reservation in the U.S. that is occupied by two tribes.

In the Second Treaty of Fort Bridger, the government provided a Bureau of Indian Affairs (BIA) agent and schools for Native children in addition to a teacher, a doctor, a carpenter, and other skilled laborers (Flynn, 2008). Under the leadership of Chief Black Coal (1878–1893), a proponent of education, land was donated to establish St. Stephen's Catholic Mission as a boarding school for Northern Arapaho children (Flynn, 2008). However, children were alienated at the mission school and prohibited from using their Native language as a form of assimilation (Hipp, 2019). In 1924, children on the WRIR attended government or mission boarding schools; however, these schools were consolidated with St. Michael's Mission School to form Mill Creek Public School District #14 (Hipp, 2019). In 1934, the Johnson O'Malley Act "provided the legal rationale for Indian control of schools on the Wind River and other reservations" (Hipp, 2019, p. 44). During the 1960s, the Shoshone and Arapaho tribal councils united to "end the government's assimilationist policies and establish self-determination in education" (Hipp, 2019, p. 50). Through the efforts of Ben Friday Jr., Alfred Redman, and many others, The Indian Self-Determination and Education Act of 1975 gave the Shoshone and Arapaho tribes power, autonomy, and sovereignty over most reservation schools (Flynn, 2008; Hipp, 2019).

Place-Based Education

John Dewey (1916) is known as the father of place-based education (PBE). He argued that students' experiences in out-of-school settings (i.e., place) should be incorporated into purposeful tasks (Long, 2009). Additionally, PBE should lead to a "multidisciplinary construct for cultural analysis…to unearth, transplant, and cross-fertilize perspectives…that can advance theory, research, and practice in education" (Gruenewald, 2003, p. 619). PBE provides Indigenous students with a foundation for learning STEM within the context of place that enriches Indigenous culture and practices by making connections through story to air, water, and land as well as ways of knowing and being (Tzou et al., 2019). Thus, PBE can be used to engage Indigenous students in socially just and anti-oppressive curriculum that values their ways of life (Stavrou & Miller, 2017).

Culturally Specific Pedagogy

According to Nieto (2002), culture is "the ever-changing values, traditions, social and political relationships, and worldview created and shared by a group of people bound together by a common history, geographic location, language, social class, and/or religion and how these are transformed by those who share them" (p. 53). "Indigenous 'cultural relevance' … refers to integration and revitalization of Indigenous histories, experiences, values, knowledge, and localized content in curriculum" (Deer, 2013, as cited in Stavrou & Miller, 2017, p. 98). However, Indigenous cultural relevance should be authentic rather than superficial and void of deficit perspectives that perpetuate racial and cultural stereotypes (Stavrou & Miller, 2017). Thus, we engaged in CSP, one of several critical pedagogies—culturally responsive pedagogy (Gay, 2010), culturally relevant pedagogy (Ladson-Billings, 2009), culturally sustaining pedagogy (Paris, 2012)—that is described as teachers' intentional use of the gestures, language, history, and cultural norms of a specific racial or ethnic group to engage students in that group in authentic learning (Leonard, 2019). CSP was taken into account when the research team chose to focus on decolonized research methods rather than quantitative methods and traditional forms of data collection, such as surveys and content tests. Furthermore, the curriculum was developed by Indigenous teachers who incorporated language, land, plants, and buffalo into STEM activities.

The Indigenous Ecological Knowledge STEM Summer Camp

Recall that the BCP was named for Bessie Coleman, the first Black and Native-American woman to receive a pilot's license. In June 2019, Jacqueline, the principal investigator (PI) of the BCP, collaborated with Native teachers to implement a one-week STEM summer camp on the WRIR. The IEK STEM summer camp, named by the instructional team, connected tribal culture to particular plant species through Indigenous Ecological Knowledge (IEK), Shoshone and Arapaho languages, and common uses for plants (Friday, 2019). The research questions were developed with decolonized research methods in mind.

Research Questions

In order to honor the Arapaho study participants, as well as their culture and traditions, the research questions were not determined *a priori* but emerged from the data and artifacts collected during the STEM camp. Two questions examine the role of language, story, culture, and place in STEM learning:

1. How did STEM instructors use Indigenous language and story to enhance STEM learning among Native-American children and youth?
2. How did STEM instructors use culture and place to engage Native-American students in learning about aviation and UAVs?

Methods

Qualitative methods were used to collect and analyze data during the IEK STEM summer camp. Central to TribalCrit are the stories and narratives that are created by Indigenous peoples about themselves, their families, and their relationships to air, land, water, plants, and animals (Bang & Marin, 2015; Brayboy & Maughan, 2009). Specifically, the case study was used as a method to tell stories about Indigenous children's and youth's STEM learning and engagement (Solórzano & Yosso, 2002). Rather than focusing on individual student responses, this study focused on Indigenous children's stories and journaling to produce counternarratives, which is a core element of CRT in general and TribalCrit specifically (Brayboy, 2005). Member checks were also implemented by sharing this chapter with an Indigenous professor of Native-American & Indigenous Studies, as well as graduate students in a course entitled, Culture, Power, and Identity in Mathematics Education. The feedback received from these individuals was used to augment this chapter.

The case study approach was used not only as a method but also the unit of analysis to examine multi-aged students' engagement in STEM (Yin, 2013). These cases were selected to highlight the process and artifacts that resulted in collaborative planning to introduce students to STEM tools, learning opportunities, and careers. The names of the locations are essential to the cases. Therefore, the WRIR is not fictitious; it is the actual place where the study occurred. All other names, with the exception of the school and town, follow Institutional Review Board (IRB) rules for anonymity as pseudonyms are used to protect student identity.

The Participants and Setting

Most of the participants in this study identified as Arapaho. One of the teachers and two staff members also identified as Arapaho. Two additional staff members, who had done significant work with Shoshone and/or Arapaho tribes, also helped to implement the program. Student participants consisted of predominantly Arapaho children and youth from six families, among others, who participated in the study. Twenty-three Arapaho students from grades four to eleven enrolled in the camp. Table 5.1 shows the number of students enrolled by grade and gender. The data show that 22% of the participants were elementary students (grades 4–6); 26% were in middle school (grades 7 & 8); and 52% were in high school. Eleven students were male and 12 were female. The inclusion of multi-aged Arapaho children and youth allowed them to learn STEM in a non-traditional, decolonized space where story and computer science merged in a highly contextualized setting (Tzou et al., 2019). The IEK STEM camp occurred at an Arapahoe school that was located on the WRIR.

Procedures

The PI, who identified as a Black female, collaborated with a middle school principal and mathematics specialist to implement the BCP with Indigenous students

TABLE 5.1 IEK STEM Camp Demographics

Grade	Male	Female
4		1
5	1	2
6	1	
7		2
8	4	
9	4	2
10		5
11	1	

in summer 2019. The PI chose not to align curriculum to traditional standards (Western values), which often perpetuate economic and cultural inequality (Aikenhead, 2001; Stavrou & Miller, 2017). Moreover, the research team avoided using Indigenous culture solely for the purpose of improving students' outcomes because this over-simplification of cultural pedagogy reinforces meritocracy and hegemonic notions that Indigenous peoples (as well as other underrepresented groups) can overcome educational inequities by simply working harder to apply themselves (Stavrou & Miller, 2017). On the contrary, Native teachers had the autonomy to develop curriculum that embraced IKS to empower students as they learned STEM (Bang et al., 2012; Tzou et al., 2019).

In the BCP study, the curriculum focused on topography, UAVs, and flight simulation in a weeklong STEM camp (see Appendix E for list of activities). Drone activities included learning basic maneuvering skills such as taking off, flying specific patterns, and landing safely. Students also learned about maps and weather, which are relevant to flight. The project leveraged IKS by using drones to collect data to learn about the land and plant species on the WRIR. Students also engaged in activities with three guest speakers, who informed them about STEM careers. The first speaker was an Arapaho artist/ecologist (female), the second was a European-American geologist (male), and the third was an African-American airline pilot (male). In addition to speakers, a field trip was planned to tie IEK to the learning environment.

Data Sources and Data Analyses

The qualitative data collected in this study consisted of artifacts that included field notes, student work samples, journal logs, and photographs. In addition to photographs, 75% of the students' journal logs included pencil drawings depicting their engagement in STEM activities. However, the resolution (less than 300 dpi) was not high enough for the images to be reproduced.

The constant comparative method (Strauss & Corbin, 1990) was used to analyze students' journal logs. However, we resisted Western ways of thinking about learning as an individual process to examine themes and patterns that emerged from Indigenous ways of knowing. Thus, we analyzed data that were collected from siblings (i.e., family clusters) and then grouped them holistically, drawing upon research findings that support the myriad ways that Indigenous families engage in informal STEM learning (Bell et al., 2009, as cited in Tzou et al., 2019).

Results

To answer the research questions, the results are presented in two parts: (a) language and story and (b) place, culture, and UAVs. Multiple sources of qualitative data are used as evidence to support the findings.

Language and Story

The story of Bessie Coleman and the excitement associated with flying drones inspired Arapaho instructors to engage in storytelling about drones (see Figure 5.1). An animal hide was used to draw an image of a drone inside a map of the WRIR. The school and nearby town of Riverton were important locations on the map. Several Native terms were introduced to make connections to Arapaho language. *Drone* was translated in Arapaho as *Beesiise' Ceebiioohut*, which is interpreted as *eyes see fly* (see Figure 5.2). The lecture notes that were associated with flying a drone were introduced to help Arapaho students make connections to their language. Warnings such as *"Watch the wind!"* and queries such as *"How did you [like] flying a drone?"* were also written in Arapaho.

In terms of story, three groups of siblings' journal logs were analyzed as a cluster for themes and patterns using the constant comparative method (Strauss & Corbin, 1990). Pseudonyms were used to assign family names to these groups of students, who described their experiences with drones and flight simulation. The following excerpts were reproduced verbatim from students' handwritten journal logs to tell their stories. While students' misspellings were not corrected, it should be noted that scientific vocabulary was not available to students during the writing process, particularly during guest speaking events.

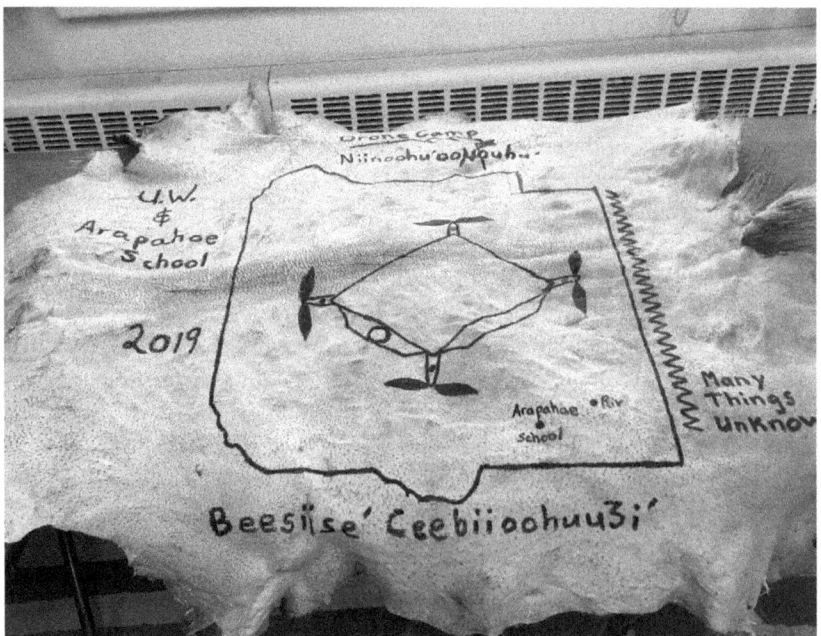

FIGURE 5.1 Animal Hide Depicting UAV on MAP of WRIR

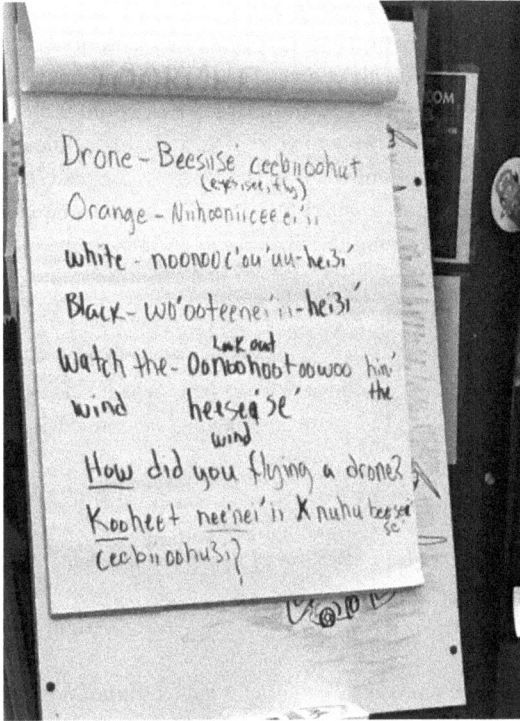

FIGURE 5.2 Lecture Notes in Arapaho

The Brownbear Siblings (Twins in Grade 10)

Today, I learned how to fly a drone. It took a few times to finally catch on. And we did a few different expeiriments that tought me a lot. Things that I didn't even know.

Lia Brownbear—Day 1

Today, I learned how to fly a plane on the computer and about what happens in each 90 day period. And the earth ain't flat! Yesterday, I learned about grasses and stuff and seen where the buffalo use to run off.

Lia Brownbear—Day 4

Okay, so today I learned how to fly a drone. I also learned how maps work. As the volasity goes up the pressure goes down.

Selena Brownbear—Day 1

The Brownbear siblings' journal logs provide a record that details their recollections of STEM experiences as well as science content. For example, Selena demonstrated an understanding of Bernoulli's law by describing the inverse relationship between velocity and pressure (*As the volasity goes up the pressure goes down.*). She also emphatically declared that ...*the earth ain't flat!* Both sisters mentioned learning how to fly a drone, and Lia commented on flight simulation, which was accomplished by using a laptop that could project any airport in the world and navigate the airplane with a joystick. Lia acknowledged learning new information that was connected to the land—*grasses and stuff*—and a sacred archeological site known as a *buffalo jump*, which demonstrated the instructors' attempt to link STEM learning to tribal places and traditions.

The Lightfoot Siblings (Grades 8 & 9)

I learned about drones. Some drones have cameras and some drones don't have cameras. Other things that drones... like little light that changes green and red. Drones can be used to look at small thing and big thing. I learned how to fly a drone.

Angie Lightfoot—Day 1

I learned that drones have other way of flying. The flit motor can change to see the other flit motor.

Angie Lightfoot—Day 2

I learned that we can make Tipi out of a map. I learned how to fly a plane and other air thing.

Angie Lightfoot—Day 3

Drone have camras, and they are good. The drone is good, and it fun and make me happy.

Eddie Lightfoot—Day 1

The drones are cool and fun, and the drone that I like is the red one. The red one is cool!

Eddie Lightfoot—Day 2

Today was fun, and it was cool. And we got to play a game [referring to flight simulation], *and that game was fun!*

Eddie Lightfoot—Day 3

The Lightfoot journal logs focused on the drone's capacity to use a camera to collect aerial images of objects and its ability to sense other drones nearby. While Angie, the female sibling, focused mostly on what the drone could do, Eddie shared his value and emotions about drones (*The drone is good, and it fun and make me happy. The drones are cool and fun. The red one is cool!*). Both siblings also reported on their experiences with the flight simulator. Again, Eddie showed that he not only valued the experience but also thought it was enjoyable (*And we got to play a game and that game was fun!*). Also, he provided text for his story about the drones and flight simulators and drew pictures. One picture showed him riding a skateboard with the drone tracking him, and the other was a portrait of him using the flight simulator. Finally, Angie's reference to the Tipi during a mapping activity revealed instructors' attempts to connect their lessons to Native-American culture.

The Bloom Siblings (Grades 5, 7, & 8)

I learned how to fly a drone and how to use maps. Also I learned how to put one together and how to hook up a phone to the drone....Also what the wind has to be and air pressure, hight it has to be.

Bryan—Day 1

I learned that the drone can follow things manuly and circle around things. Also I learned how to track mountains on a road map. I learned how to record things and how to take different shots. I learned lots of different controls with the drones.

Bryan—Day 2

I learned how to use the intelligent mode on the drone. Also I learned how to play the flight simulator. And I saw pilots today that let use play the flight simulator.

Bryan—Day 3

What I learned was how to control a drone. Also how to do simple tricks. ...they get some of the best pictures and videos. Also that they can stand strong winds up to 20 mph.

Stanley—Day 1

Today, I learned how to fly and make a drone do stuff. Like make an eight and a square. We also learned what things are called on the drone. The most hardest thing to learn was the eight. The easyest to learn was how to fly. You could put your phone to the controller. It make it take pictures and fly it with your phone.

Amber—Day 1

I had fun recording the river, then everyone seeing it. My brother Stanley didn't come today. Bryan came, and me and him made a bet. So I learned that the drones can go in a perfect circle by just pressing a button.

Amber—Day 2

> *Today, I learned that if you get to close to someone else flying a drone, your drone could be disconnected, or if your drone gets to far away from you. We were recording one of the teachers run around the school. The other teachers thought I was using the phone to track them, but they were surprised when they found out I was flying it. We also played on the computer. It was like practicing flying the drones. Yesterday I found a rock that looked cool! The rock looked like a sandwich.*
>
> Amber—Day 3

Three siblings from the Bloom family participated in the IEK STEM summer camp. Like the other siblings, they focused on drones and flight simulation as they told stories about their experiences through journaling. All of the siblings illustrated their stories with drawings. However, the narratives placed focus on weather conditions (e.g., wind and air pressure) and what the drones were capable of doing (e.g., follow/track, fly in a perfect circle, make a figure eight or a square, take pictures and videos, etc.). Bryan and Amber also reported learning how to put a phone in the controller, which can be used to guide the drone. Additionally, Amber mentioned recording the *river* and finding a *rock* at Beaver Rim (*Yesterday I found a rock that looked cool!*). Bryan mentioned the importance of using a road map to locate *mountains*. Thus, two of the three siblings made connections to the land and water, which are significant in Indigenous culture. Only Bryan mentioned the flight simulator and the guest speaker, who was an airline pilot. Lastly, Amber provided evidence that learning was communal as three drawings in her journal logs illustrated group learning in the classroom, flying the drone around the basketball hoop, and tracking her brother, Bryan, with the drone.

It is evident from the data presented earlier that the instructors laid a strong foundation for Indigenous students to use language and story (i.e., text and drawings) to describe their learning during the weeklong camp. Students engaged in communal learning as they drew upon Native-American culture (Tipi and geographic location) and emerging technology (i.e., smart phones and drones) to learn STEM in an informal learning environment.

Place, Culture, and UAVs

During the IEK STEM summer camp, instructors were purposeful in their use of place and culture to make the lessons relevant and interesting to Indigenous students. One of the lessons that stood out related to understanding topographical maps. A second lesson took place during the day trip to Beaver Rim, where students gained IEK in real time using drones to collect aerial footage.

Place-Based Education

The topography lesson was relevant to place because the WRIR is located in the Wind River Basin, which includes part of the Wind River Mountain Range,

FIGURE 5.3   Raise Relief Map of WRIR

Owl Creek Mountains, and Absaroka Mountains. The lead instructor in this lesson showed the students where they lived on a 3D map known as a *raised relief map* (see Figure 5.3).

After connecting the raised relief map to place, students learned how 2D maps can be used to show different levels of elevation with contour lines. The lesson involved using potatoes of different sizes and shapes to resemble a mountain. Students drew the contour lines to show what the mountain might look like from an aerial view. This activity enhanced students' visualization skills as they tried to capture a 3D image in a 2D space. Figure 5.4 shows one of the students using the potatoes to make a contour map.

All of the students' journals were analyzed for themes and patterns related to STEM content learning. Twenty students made daily entries in their journals. Forty percent made references to maps, and only one comment was negative. Two students expanded their learning to include Geographic Information Systems (GIS). A few of their comments are presented as follow:

> *I searched…my house on Google images.*
>
> (female, Grade 5)

> *The maps were a little boring….*
>
> (female, Grade 7)

> *We looked at Google Earth on the Chromebooks.*
>
> (male, Grade 8)

FIGURE 5.4 Student Working on Contour Map

I learned how maps work.

(male, Grade 9)

The instructors also made connections to Indigenous culture and place by organizing a day trip to Beaver Rim, which is a cliff area with an elevation of 7,162 feet, to collect data and to see the buffalo jump. The hilly area and valley below Beaver Rim was an ideal location to drive buffalo. As the buffalo fell from the cliffs, they were immobilized and killed for food and clothing. More often than not, Native Americans camped near buffalo jumps leaving artifacts that depicted everyday life behind. For some Arapaho children, this was a novel experience.

The trip took more than an hour from Riverton and required driving on dirt roads for the most part. A geologist accompanied the group, which included the PI, three Arapaho teachers/staff, two additional staff, two volunteers from the Teton Science Schools, and 21 Arapaho children/youth. The group enjoyed lunches that were packed for the trip and listened to the geologist, who told them about the plants and minerals that could be found on Beaver Rim. After learning about plants and minerals, drones were flown to collect aerial footage of plants, land, and the buffalo jump (see Figure 5.5). All of the students had the chance to fly drones that were both large and small (e.g., Phantom, Mavik Pro, Spark drones by DJI).

Analyses of students' journal logs revealed that three Arapaho females made comments about the field trip to Beaver Rim. However, the limited number of comments should not be interpreted as an indication that the experience did not resonate with the students as some Indigenous students are reluctant to talk about their culture (Leonard et al., 2018). As a consequence of systemic racism,

FIGURE 5.5 Indigenous Student Reflects at Buffalo Jump

colonization, and assimilation, Native students may not be aware of their languages and cultures (Stavrou & Miller, 2017). Nevertheless, the girls' comments reveal that instructors' attempts to make connections to culture and place were authentic and impacted students' learning about plants, animals, weather, and tribal ritual.

> *It's windy at Beaver Rim.*
>
> (female, Grade 9)

> *I learned about grasses and stuff. And seen where the buffalo used to run off.*
>
> (female, Grade 10)

> *I learned that a buffalo wallow is a circle and hoof prints belonging to buffalos in a herd. But a buffalo jump is where our ancestors tricked buffalo, and buffalos end up jumping off the cliff to kill them for food. We also seen a wild horse run towards us.*
>
> (female, Grade 10)

Noticeably, the Arapaho students were unafraid to venture out on the rim. Field notes and photographs collected by the PI revealed the students had a quiet reverence for the buffalo jump.

UAVs/Drones

The main focus of the IEK STEM summer camp was learning about drones and using them as an introduction to exploring careers in aviation. All 20 students who completed journal logs for one or more days reported learning about drones. Several comments focused on learning maneuvers, such as take-off and landing, use of cameras and videos, optimal flying conditions, flight patterns (e.g., circles, squares, 180s, figure eights, and so on). Sixty percent of the journals included mentions of the drone's tracking ability. Some of the more salient comments are presented subsequently:

I learned that drones can follow people, and I learned that you can create a film by using a drone. You can also make a photo collage.

(female, Grade 4)

I had fun recording the river, then everyone seeing it.

(female, Grade 7)

Today, I learned drones can follow people, and I seen a drone video with…on a skateboard. He falls, but he's okay I think. Drones can fly themselves home when they are about to die.

(male, Grade 9)

I learned that you can track objects…that drones have manuvers and can take pics or videos. I also got to fly the tiny and big drone a lil today. Flying the tiny drone was more easier though. While I was flying a drone, I took group pics. LOL!

(female 1, Grade 10)

I learned today how to fly a drone. I learned how to do circles and 180s and how to land and start to fly. We watched the view from the drone. Me…and…took pictures together from the drone. I saw…fall off of a skateboard and the drone caught it on video.

(female 2, Grade 10)

Two themes related to Indigenous identity emerged from these four students' journal logs. The first had to do with *presencing* as these Arapaho students used drones to document their lives with photography and video. Two of the high school girls specifically mentioned using the drone to take group photos: *While I was flying a drone, I took group pics; Me…and…took pictures together from the drone.* The photos were more than just snapshots; they symbolized sisterhood and community. Thus, drones allowed Arapaho students to capture not only their presence during a STEM activity but also the kinship nature of their identity based on relationships (Borden, 2011; Lipka et al., 1998; Tzou et al., 2019).

The second theme was the reference to skateboarding: *I seen a drone video with…on a skateboard*; and *I saw…fall off of a skateboard and the drone caught it on video.* Field notes revealed that several boys brought skateboards to camp. While an ordinary pastime for many young men, some of the older males in this study identified with the recreational sport of skateboarding. Two journal logs implied that skateboarding can be treacherous; however, empathy emerged in one of the journals: *He falls, but he's okay I think.* Skateboarding was part of these Arapaho students' culture, particularly for males.

Finally, some students connected the drones to STEM careers and specifically referenced the guest speaker who was a United Airlines pilot. Analyses of journal logs revealed students' learned about aviation. Additionally, they reflected on flying airplanes or drones as a career path.

> *Today, I found out that you can make money by flying drones. I hope to learn more about drones at drone camp.*
>
> (female, Grade 4)

> *...I saw pilots today that let us play the flight simulator.*
>
> (male, Grade 5)

> *I learned that when you're a pilot, in four years you make $1 million.*
>
> (female, Grade 5)

> *Today, was fun because a guy came in and told us about airplanes, latitude and longitude, hours and earth time, speed, that north, south, west, and east are the directions.*
>
> (male, Grade 9)

While no students actually stated that they wanted to be a pilot or fly drones as a career, the foregoing comments are promising. The youngest student participant in the study stated that she wanted to learn more about drones: *I hope to learn more about drones at drone camp.* One student associated the pilot with playing the flight simulator, while another associated the pilot's presence with having a *fun* day. These comments speak to the importance of having role models from underrepresented backgrounds to share their experiences with students (see Chapter 4 of this volume). Taken together, the qualitative data suggest that the IEK STEM summer camp was successful in meeting its target goal. Arapaho students learned how to fly drones and used them purposefully to tell stories about their cultural and social identities. Instructors made explicit connections to place and culture while engaging students in meaningful STEM activities to broaden Indigenous students' awareness of STEM and STEM-related career paths.

Discussion

The results of this study of Indigenous students, who participated in the IEK STEM summer camp, reveal three important findings. The first and most remarkable is that empowering Native teachers and staff with the autonomy to develop culturally responsive curriculum for students yielded high interest and retention among Arapaho children and youth that led to rich STEM content learning. The teachers used words from the Arapaho language to describe drones, which enthralled children and youth in the IEK summer camp. Every single journal mentioned drones, some with illustrations depicting how students used drones purposefully. For example, students used the drones to collect photographs and videos that captured kinship relationships, skateboarding, landscapes, and sacred spaces. Students then created narratives about drones and described how they were used along with original drawings in their journals.

In terms of STEM learning, students demonstrated an understanding of Bernoulli's law, the principles of flight, and the importance of knowing about weather, cardinal directions, wind, and speed to fly an airplane or a drone. By using a decolonized approach that focused on the identities and assets of the Northern Arapaho peoples, children made use of *presencing* in their journals with narratives that usually began with "*Today, I learned….*" to described their STEM learning and engagement (Tzou et al., 2019). Students also engaged in *verbification* (i.e., following, tracking, fly home, etc.) to describe drone activity (Borden, 2011). Thus, three or more TribalCrit tenets were evident in the study reported here (e.g., right to sovereignty, autonomy, self-determination, and self-identification; viewing culture, knowledge and power through an Indigenous lens; and recognizing stories as relevant data) (Brayboy, 2005).

Second, the instructors' use of place and culture helped students to learn how drones could be used to tell rich stories about their communities. PBE allowed Arapaho children and youth to engage in activities that blended cultural heritage, traditions, and landscapes (Gruenewald, 2003). CSP was used to promote the Arapaho language and history as well as to engage Arapaho students in authentic STEM activities (Leonard, 2019). Families that consisted of multi-aged siblings engaged in computational participation as they worked together to capture drone images of water, land, and plants. Young women gathered to document their kinship by taking photographs of themselves with a drone to portray their STEM identity, and young men celebrated their skateboarding identity with drone tracking and video (Bang et al., 2012; Tzou et al., 2019).

Finally, in terms of broadening participation with Indigenous peoples in STEM, students acknowledged that airline and drone pilots can earn large incomes. However, high salaries are not enough to attract members of every ethnic group to the pursuit of STEM. While no racial/ethnic group is monolithic, Indigenous people often prefer relational rather than transactional relationships (Borden, 2011; Brayboy & Maughan, 2009; Minero, 2019). In particular, family, culture, traditions, place, and community are highly valued (Florentine, 2019; Minero, 2019). Connecting STEM learning to place and culture, working with role models from Native backgrounds, showing how STEM (and art) can be used to enhance the lives of tribal communities, and recognizing the value of working remotely from the reservation can change Indigenous peoples' perceptions about STEM and STEM-related careers (Minero, 2019; Tzou et al., 2019).

Summary

This case study used counternarratives to describe Arapaho students' experiences with drones and flight simulation as the context for STEM learning. The research team initially saw flight as a way of using place and culture as springboards to learn about science and the physics of flight as tools to excite students about pursuing careers in STEM fields. However, the instructional team at the Arapahoe school took these ideas further by using the drones to engage their students in meaningful site-specific data collection and sociocultural identity as they explored their past and looked toward their future. The instructional teams' knowledge and experiences were critical in designing learning experiences for the IEK STEM camp that were place-based and culturally specific at a finer scale. These critical pedagogies enhanced STEM learning for student participants in the BCP. Continued exposure to projects of this nature, as well as IEK STEM curriculum and critical pedagogy, may encourage Indigenous students to pursue careers in STEM. Future research endeavors should include tapping Indigenous STEM professionals to serve as guest speakers not only to learn what influenced them to work in a STEM field but also to use that knowledge to broaden the STEM participation of children and youth from similar backgrounds.

In conclusion, "Native Americans have always been scientists, innovators, and engineers" (Minero, 2019, p. 4). Therefore, it is imperative for researchers of large-scale projects, such as the BCP, to use decolonized research principles to build upon Indigenous ways of knowing to empower students to reach their full potential—not only as STEM learners but as STEM knowledge producers as well.

Note

1 Tenets are not listed in the same order as they are presented in Brayboy (2005).

References

Aikenhead, G. S. (2001). Students' ease in crossing cultural borders in school science. *Science Education*, *85*(2), 180–188.

Anderson, C. R. (2019). "Critical what what?": Critical race theory and mathematics education. In J. Davis & C. C. Jett (Eds.), *Critical race theory in mathematics education* (pp. 18–31). Routledge.

Bang, M., & Marin, A. (2015). Nature–culture constructs in science learning: Human/nonhuman agency and intentionality. *Journal of Research in Science Teaching*, *52*(4), 530–544. doi:10.1002/tea.21204

Bang, M., Warren, B., Rosebery, A. S., & Medlin, D. (2012). Desettling expectations in science education. *Human Development*, *55*(5–6), 302–318. doi:10.1159/000345322

Bell, D. (1987). *And we are still not saved: The elusive quest for racial justice.* Basic Books.

Borden, L. L. (2011). The 'verbification' of mathematics: Using the grammatical structures of Mi'kmaq to support student learning. *For the Learning of Mathematics*, *31*(3), 8–13.

Bracey, J. (2013). The culture of learning environments: Black student engagement and cognition in math. In J. Leonard & D. B. Martin (Eds.), *The brilliance of Black children in mathematics: Beyond the numbers and toward new discourse* (pp. 171–194). Information Age Publishing.

Brayboy, B. M. J. (2005). Toward a tribal critical race theory in education. *The Urban Review, 37*(5), 425–446.

Brayboy, B. M. J., & Maughan, E. (2009). Indigenous knowledges and the story of the bean. *Harvard Educational Review, 79*, 1–21. doi:10.17763/haer.79.1.l0u6435086352229

Bureau of Labor Statistics, U.S. Department of Labor. (2016). *Occupational outlook handbook.* http://www.bls.gov/ooh/computer-and-informationtechnology/computer-and-information-research-scientists.htm.

Crenshaw, K. W. (1988). Race, reform, and retrenchment: Transformation and legitimation in antidiscrimination law. *Harvard Law Review, 101*(7), 1331–1387.

Dewey, J. (1916). *Democracy and education: An introduction to the philosophy of education.* Macmillan.

Florentine, S. (2019). Diverse-it: Intel's Native coders program offers pathway to STEM for Native American students. *CIO,* n.p. https://www.cio.com/article/3340325/intel-snative-coders-program-offers-pathway-to-stem-for-native-american-students.html

Flynn, J. (2008). *Tribal government: Wind River Reservation.* Mortimore Publishing.

Friday, C. (2019). *Linking Native American culture with the rangeland resources on the Wind River Indian Reservation in Wyoming.* Unpublished thesis: University of Wyoming.

Gay, G. (2010). *Culturally responsive teaching: Theory, practice and research* (2nd ed.). Teachers College Press.

Gruenewald, D. A. (2003). Foundations of place: A multidisciplinary framework for place-conscious education. *American Educational Research Journal, 40*(3), 619–654.

Hipp, M. L. (2019). *Sovereign schools: How Shoshones and Arapahos created a high school on the Wind River Reservation.* University of Nebraska Press. doi:10.2307/j.ctvd1cb4q

Kafai, Y. B., & Burke, Q. (2014). *Connected code: Why children need to learn programming.* The MIT Press.

Ladson-Billings, G. (2009). *The dreamkeepers: Successful teachers of African American children* (2nd ed.). Jossey-Bass.

Leonard, J. (2019). *Culturally specific pedagogy in the mathematics classroom: Strategies for teachers and students* (2nd ed.). Routledge.

Leonard, J., Buss, A., Gamboa, R., Mitchell, M., Fashola, O. S., Hubert, T., & Almughyirah, S. (2016). Using robotics and game design to enhance children's STEM attitudes and computational thinking skills. *Journal of Science Education and Technology, 28*(6), 860–876. doi:10.1007/s10956-016-9628-2

Leonard, J., Mitchell, M., Barnes-Johnson, J., Unertl, A., Outka-Hill, J., Robinson, R., & Hester-Croff, C. (2018). Preparing teachers to engage rural students in computational thinking through robotics, game design, and culturally responsive teaching. *Journal of Teacher Education, 69*(4), 386–407. doi:10.1177/0022487117732317

Lipka, J., Mohatt, G. V., & the Ciulistet Group. (1998). *Transforming the culture of schools: Yup'ik Eskimo examples.* Lawrence Erlbaum Associates.

Long, V. M. (2009, September/October). Adding "place" value to your mathematics instruction. *Connect Magazine,* 10–12.

Minero, E. (2019). Building a STEM pathway for Native students. *Edutopia,* n.p. https://www.edutopia.org/article/building-stem-pathway-native-students

National Science Foundation (2017). *Women, minorities, and persons with disabilities in science and engineering.* National Science Foundation, National Center for Science and Engineering Statistics, National Survey of College Graduates 2015. https://www.nsf.gov/statistics/2017/nsf17310/static/data/tab9-19.pdf

Nieto, S. (2002). *Language, culture, and teaching: Critical perspectives for a new century.* Lawrence Erlbaum Associates.

Paris, D. (2012). Culturally sustaining pedagogy: A needed change in stance, terminology, and practice. *Educational Researcher, 41*(3), 93–97. doi:10.3102/0013189X12441244.

Solórzano, D. G., & Yosso, T. J. (2002). Critical race methodology: Counter-Story telling as an analytical framework for education. *Qualitative Inquiry, 8*(1), 23–44.

Stavrou, S. G., & Miller, D. (2017). Miscalculations: Decolonizing and anti-oppressive discourses and Indigenous mathematics education. *Canadian Journal in Education/La revue Canadienne de l'apprentissage et de la technologie, 40*(3), 92–122.

Strauss, A. L., & Corbin, J. (1990). *Basics of qualitative research: Grounded theory procedures and techniques.* Sage.

Tzou, C., Suárez, E., Bell, P., LaBonte, D., Starks, E., & Bang, M. (2019). Storywork in STEM-Art: Making, materiality and robotic within everyday activities of Indigenous presence and resurgence. *Cognition and Instruction, 37*(3), 306–326. https://doi.org/10.1080/07370008.2019.1624547

Varma, R. (2009). Attracting Native Americans to computing. *Communications of the ACM, 52*(8), 137–140. doi:10.1145/1536616.1536650

Yin, R. K. (2013). *Case study research: Design and methods.* Sage.

Yosso, T. J. (2006). *Critical race counterstories along the Chicana/Chicano educational pipeline.* Routledge.

6

PROFESSIONAL DEVELOPMENT THAT FOSTERS COMPUTATIONAL THINKING AND HIGH-QUALITY TEACHING FOR STUDENTS OF COLOR

Roni Ellington and Jacqueline Leonard

I believe that kids are kids. I serve in an area that is predominantly Black. It's still mixed and, being in a predominantly Black area, a lot of misconceptions and stereotypes happen with our students. I want my students to be able to just do everything that I know they can do. There is no deficit in our children, so I want my kids to do everything that I want them to do, and they honestly always succeed. In my classroom, we always say first everything is possible and there aren't any limits … [In] the ITEST program, when we worked on video gaming, I had the opportunity to see … how things [were] placed together, and we did coding; it was great … to see my students were excited about this; that's what made me even more excited as well.

We are going to be flying drones at the end of the school year. [My students] are also going to know their coordinate geometry, and they are going to learn about probability. They are also going to be able to understand number and concepts and their conversion measurements. I am that excited, I cannot wait to get back to write a lesson plan about this

Ms. Davis, Sixth-Grade Teacher
August 2018

This vignette reflects the experiences of a Black female teacher who participated in a professional development experience through the Bessie Coleman Project (BCP), an out-of-school project designed to promote computational thinking (CT) among elementary and middle school students through the use of computer modeling, 3D printing, flight simulation, and drones. This teacher, who graduated with a bachelor's degree in mathematics, obtained a Master's degree in curriculum, and had 17 years of teaching experience (at the time of this interview), exhibited the kind of perspectives of Black children in mathematics that serve as the foundation of effective teaching, particularly instruction that focuses on CT. The vignette also reflects this teacher's positive beliefs and attitudes about her students, which emphasizes what they can do and the brilliance that they bring to their mathematics experiences (Gholson, 2017). She was also able to reflect on the misconceptions and stereotypes that are projected onto students of color, and therefore, positions herself as a gateway (as opposed to a gatekeeper) for her students to fully participate in STEM. She saw how the experiences she had in the out-of-school learning environment could be applied to topics in the formal learning environment. More importantly, she demonstrated a level of enthusiasm and excitement for how these informal learning experiences were beneficial to her mathematics class and how these experiences could be used to develop her students' problem-solving and CT skills.

The vignette also reflects how this teacher encouraged her students to engage in coding, a form of CT, as an approach to problem solving that incorporates thought processes, such as abstraction, decomposition, algorithmic design, evaluation, and generalizations (Selby & Woollard, 2013). Furthermore, she saw herself as a part of the community of learners and embodied the tenets of computational participation (Kafai, 2016; Kafai & Burke, 2017). Having teachers who help foster CT/participation is critical for teaching students of color.

In this chapter, we provide a discussion of how to frame professional development, also known as PD, which will help teachers plan and implement instruction that can foster CT and participation for students of color. We begin with a discussion of the extant literature that illuminates how CT and participation can be incorporated into PD experiences—specifically, how out-of- school programs can provide real-world contexts that support teachers' self-efficacy, CT competencies, and development of positive dispositions (Pérez, 2018). Next, we explore how to integrate equity and culturally responsive approaches into PD to support teachers in adopting assets-based narratives and approaches that provide the foundation for teaching CT to students of color. The preliminary results of this project highlight how the PD impacted participants' teaching efficacy beliefs and attitudes toward STEM. In addition, we report on results obtained from the Dimensions of Success (DoS), a tool used to rate teachers' enactment of STEM practices and their interactions with students (Shah et al., 2014). Data collected via the DoS instrument were used to illustrate how teachers' practices were impacted by their experiences in the project, and we connect those observations to participants' earlier exposure to in-depth PD. We conclude the chapter with a summary of the PD implemented as a part of the BCP, which aimed to provide underrepresented students with out-of-school learning opportunities, such as 3D printing and drones to motivate awareness of, interest in, and preparation for careers in STEM-related fields.

Computational Thinking, Participation, and Dispositions in Professional Development

A growing body of research suggests that infusing CT into K-12 STEM instruction will support the skills and competencies needed by students for future STEM careers (Foster, 2006). Computational thinking, broadly defined as formulating and solving problems using computational steps and algorithms (Aho, 2012) (a) helps students develop reasoning skills that are needed to use and analyze big data (Hunsaker, 2020); (b) assess and solve unstructured problems (Lee et al., 2014); (c) design and use computational models (Kafura et al., 2015); (d) decompose tasks (Dietz et al., 2019); and (e) generalize and transfer problem-solving processes to novel situations, which are skills that are required in the 21st-century workplace (Grover & Pea, 2013). In addition to preparing students for careers, infusing CT into instruction has the potential to increase student achievement and persistence, specifically for minority students (Tran, 2018). Therefore, increased attention to CT has been advocated by scholars and educators to help students expand their understanding of mathematics and STEM concepts (Grover & Pea, 2013; Lye & Koh, 2014; Weintrop et al., 2016; Wing, 2006).

Not only is there a focus on building students' computational skills, there is also increased interest in promoting students' computational participation. In fact, some scholars have argued for a move from CT toward computational participation, which emphasizes the notion that programmed projects such as games, stories, and art are not simply "objects-to-think-with" but also "objects-to-share-with" that connect us to and with others (Kafai & Burke, 2014, p. 109). Hence, computational participation requires solving problems with others, designing intuitive systems for and with others, and learning about the cultural and social nature of human behavior through the concepts, practices, and perspectives of computer science (Brennan & Resnick, 2012). Having children program applications, work in groups, and remix code moves CT from an individual effort to a community effort and encourages formal and informal collaboration (Kafai & Burke, 2014). By making CT a community effort, students and teachers can leverage their computational competencies to positively impact themselves, others, and their community.

Literature focused on understanding the types of dispositions that will allow students to make use of their CT skills and engage in computational participation is emerging (Brennan & Resnick, 2012; Pérez, 2018). As Pérez (2018) asserted, serious consideration to the development of computational dispositions is critical to the connective possibilities of CT practices. "Dispositions reflect how inclination, sensitivity and ability to interact in a learner's thoughts and actions in a given situation and the development of positive dispositions has been shown to impact the quality of learning opportunities afforded to students" (Dweck, 2006, as cited in Pérez, 2018, p. 435). Engaging learners in CT requires recognizing and cultivating dispositions that include tolerance for ambiguity, persistence on difficult problems, and collaboration with others to achieve a goal that promotes the implementation of problem-solving practices.

Growing interest in CT in K-12 education has resulted in an increasing number of research studies; however, little attention has been paid to pre-service and in-service teacher training (García-Peñalvo et al., 2016; Grover & Pea, 2013; Yadav et al., 2017). Although research in this area is scant, studies are emerging that highlight the kinds of PD that supports teachers in their understanding of CT as well as their implementation of instructional practices that promote CT (Yadav et al., 2016). This literature suggests that effective PD should help teachers develop their own CT, provide real-world collaborative experiences that connect conceptual understanding to real-world problems, and supports them in translating their PD experiences to instructional activities that they can use in their classrooms (Asghar et al., 2012; Avery & Reeve, 2013).

Scholars (Brennan & Resnick, 2012) suggest that PD on CT should help teachers learn and apply core concepts and practices that include the following: (a) understanding of sequences, loops, events, conditionals, and operators; (b) assessing different approaches to solutions to a problem; (c) generalizing and transferring the problem-solving process to various situations; and (d) understanding how data originates from diverse sources. CT focused PD, as promoted in the literature (Killen et al., 2020; Kong et al., 2020), should also support teachers in the following: (a) understanding techniques employed to cope with quality data; (b) organizing, describing, and managing data; (c) understanding the diverse motivations and barriers associated with data sharing, production, and consumption; and (d) understanding and explaining the role of modeling in problem solving and algorithmic thinking (Aho, 2012; Wing, 2008).

Li et al. (2019) conducted a study that explored the influence of a blended PD model on teachers' learning of content and pedagogical knowledge related to CT. Results showed that blended PD enhanced teachers' understanding of CT, learner-centered multidisciplinary approaches to incorporating CT into their lessons, and differentiated learning strategies. Furthermore, the findings revealed that teachers' integration of CT into their practices reflected the following: (a) teachers' background knowledge; (b) thoughtful consideration of student interests; and (c) the content taught (Li et al., 2019). In another study, Ketelhut et al. (2019) examined teacher changes following PD that integrated CT into elementary school science. Professional development consisted of two parts: an initial half-day Saturday session and seven 90-minute afterschool sessions (i.e., inquiry groups) that included mentor teachers and pre-service teachers. The results of the study revealed that the PD experience shaped teachers' beliefs about CT and how it could be incorporated into their classes. The potential for CT-infused activities to be motivating, interesting, and relatable to students offered opportunities to use best practices in science teaching that included: (a) encouraging student collaboration; (b) engaging in problem solving; and (c) making real-world and interdisciplinary connections (Ketelhut et al., 2019). Notably, teachers grew to believe that integrating CT engaged student learners from all backgrounds.

Some scholars have suggested that out-of-school programs can be effective in helping students to develop CT skills, engage in computational participation, and cultivate positive computational dispositions (Bhaduri et al., 2018; National Research Council, 2011). For example, Buss and Gamboa (2017) conducted a study centered on the program entitled Visualization Basics: Using Gaming to Improve Computational Thinking, also known as uGame-iCompute. This was a program that provided opportunities for teachers and middle school students to participate in afterschool clubs that focused on game design and LEGO robotics in an effort to enhance the participants' CT skills. To help instructors develop proficiency with the tools (i.e., gaming platform and robots) and strategies to develop CT skills for themselves and their students, researchers developed an eight-week PD course.

In Year 3 of the uGame-iCompute study, 28 teachers, including Ms. Davis in the vignette given earlier, enrolled in the PD course that met 2.5 hours per week for a total of 20 contact hours. The course focused on a set of challenges that involved game design and robotics. The results of the project revealed that instructors' understanding of and attitudes toward CT improved. Results also revealed that teachers provided students with a mix of open-ended and exploratory activities that encouraged them and supported their efforts at problem solving. However, Buss and Gamboa noted some program limitations that centered on the length of the course. While instructors, including Ms. Davis, clearly benefited from the PD, the eight-week course was not long enough to familiarize all teachers with CT. Some teachers did not have the time to complete the CT tasks for themselves or delve into how to translate their experiences into classroom instruction. Despite these limitations, the results of the uGame-iCompute study suggest that afterschool programs can be effective in enhancing teachers' and students' CT skills, and they can increase teachers' comfort with computing and instructional practices that lead to CT.

These studies show that PD that focuses on CT should include opportunities for teachers to actively engage in CT/participation over an extended period of time. In addition, designing PD that used a blended learning model promoted collaboration, allowed teachers to develop their own CT skills, and enhanced teachers' positive dispositions regarding CT as well as their use of strategies to integrate activities that fostered CT in their students. These PD activities were structured but provided enough flexibility to encourage teachers to be creative about how they could integrate what they had learned into their instructional practices.

Culturally Responsive Equity-Based Professional Development

There has been a growing body of work that highlights how STEM pre-service and in-service PD can be designed to achieve the goal of creating reflective, critically conscious teachers, who are empowered to address the needs of students of color as STEM learners and enact pedagogical practices that liberate these students (Joseph et al., 2015). Hence, preparing teachers to provide effective CT instruction to students of color must first be grounded in an equity-centered culturally responsive framework. If teachers hold narratives about students of color that focus on what they believe they cannot do, it may impede their willingness to implement these activities in their classes (Bower et al., 2017; Darling-Hammond et al., 2019). STEM teachers who teach historically-excluded students must be reflective practitioners who are willing to reflect not only on their teaching practices but also on who they are, their beliefs and perspectives, and the various biases they bring into the classroom (Gay & Kirkland, 2003). Specifically, professional developers must find ways to challenge teachers to interrogate dominant deficit-based metanarratives about children of color and build counternarratives that affirm the collective intellectual ability and identity of students of color (Delpit, 2012; Martin, 2009). This is particularly important in CT focused PD, given that many teachers may hold views that CT is for the *gifted* or *mathematically talented.*

Battey and Franke (2015) provided a conceptual framework for PD that integrated mathematics that was specifically designed for urban mathematics teachers. Although the PD was not focused on CT, Battey and Franke offered a framework that can be applied to CT. Using vignettes that highlight deficit stories about children of color, Battey and Franke employed an approach to PD that integrated equity and mathematics that drew heavily on the idea of metanarratives, specifically those that reflected the participation of students of color in mathematics (Jacobs et al., 2007). The findings suggest that the PD sessions helped teachers reframe deficit metanarratives. Professional development that integrates equity along with mathematical content has the potential to improve the educational outcomes of students. This is particularly important for teachers, who attempt to infuse CT/participation into their instruction because it requires innovation and disruption of traditional models of teaching and learning.

In addition to countering negative metanarratives about students of color, several scholars argue that both pre-service and in-service teachers must develop a type of critical consciousness that allows them to interrogate their own beliefs and attitudes about children of color and adopt more asset-based perspectives of the children they teach (Battey, 2013; Celedón-Pattichis et al., 2018; Martin, 2012). Teachers should develop an *engaged pedagogy* that requires a reflective stance that involves interrogating one's location and the identifications and allegiances that inform one's life (hooks, 1994), which serves as the foundation of any pedagogy designed to improve the learning outcomes of students of color. This is particularly important for teachers, who are integrating CT in their practice because this integration can lead teachers to experience their own cognitive struggles, which can reinforce limiting beliefs about their students' ability to engage in CT. Specifically, PD that does not allow teachers to interrogate and transform their negative beliefs and perspectives about students of color can lead them to abandon efforts to infuse CT into their instruction. Hence, scholars recommend that PD promotes self-awareness, requires teachers to engage in activities that require cognitive struggle similar to that their students face, and allows them to grapple with their own beliefs and attitudes about students (Buss & Gamboa, 2017).

By creating PD opportunities that integrate effective practices in CT/participation, teachers will be more empowered to address the myriad of challenges they face in teaching students of color. Moreover, they will become better equipped to build the kinds of relationships with students of color that reflect a culture of care in order to nurture their students' STEM learning, empower them to be resilient learners, and create classroom environments where students feel a sense of belonging. Furthermore, PD should address not only the struggles teachers face with teaching STEM content but also other issues, such as classroom management, student disengagement, and other academic issues to yield more positive learning outcomes.

Teachers' Identity, Background, and Characteristics in Professional Development

> Although it seems obvious that personal experiences contribute to the formulation of our beliefs and the way we act, as we know from our many experiences participating in and facilitating PD, facilitators do not always take into account the ways in which teachers' lived experiences contribute to their beliefs about and the way they practice teaching.
>
> (Wager & Foote, 2013, p. 24)

In some situations, a teacher may need to confront prior experiences before making a change in practice. Thus, it is useful for teachers to examine past experiences and consider the ways those experiences have contributed to who they are, and consequently, to how they teach. Therefore, facilitators of PD should examine teachers' past experiences to understand how they engage with PD, specifically the ways in which their personal identities and lived experience shape not only their views about content, such as mathematics, but also about the students they teach (Kong et al., 2020). Addressing teachers' identities and lived experiences in PD that focuses on CT/participation is important because of what is required of teachers to integrate CT into their work.

Research suggests that teachers' background and experience are important because their content knowledge, beliefs, and philosophy influence how they experience PD and facilitate instruction. Mitchell (1998) contended that "minority teachers are particularly adept at motivating and engaging minority students because they often bring knowledge of student background to the classroom that enhances students' educational experience" (p. 105). Moreover, Clark et al. (2013) argue that "the work and role of the African-American mathematics teacher take on particular features and serve purposes beyond that of facilitator of mathematics learning for students" (p. 5). Thus, our assumption is that teachers are more likely to engage predominantly Black students in spatial reasoning and CT/participation in culturally specific ways when they are sensitive to the needs of these students and see their role as an amalgam of teacher and agent of empowerment (Ellington, 2019). Therefore, the opening vignette provides a snapshot of how Ms. Davis's background as an African-American woman contributed to her understanding of and willingness to combat negative stereotypes. Her background and identity as an African-American teacher influenced how she experienced the PD. As a result, she was willing to implement effective CT in two different afterschool programs (i.e., uGame-iCompute and the BCP), which will be discussed in more detail later in the chapter.

Teacher Efficacy and Beliefs

In addition to drawing on teachers' background knowledge about students, it is also critical that PD consider teachers' beliefs, self-efficacy, and outcome expectancy related to CT. Self-efficacy is defined as "one's perceived capabilities for learning or performing a task" (Schunk, 2020, pp. 11–12). Bandura (1997) described self-efficacy in two parts: (a) beliefs about one's capabilities and (b) agency related to bringing about outcomes. Teachers with high self-efficacy are more likely to develop challenging activities, help students succeed, and persevere with students who have problems learning (Schunk, 1991). Research suggests that teachers with high self-efficacy are also more likely to exhibit high-quality pedagogical skills (Bandura, 1997; Siwatu, 2007).

As it relates to teaching, developing a strong sense of self-efficacy is important to implementing a new pedagogical approach, such as CT (Leonard et al., 2018; Yadav et al., 2016). Yang (2019) found that teacher self-efficacy can be enhanced by learning about the experiences of teachers participating in PD. Barni et al. (2019) found that teachers' self-efficacy was a major factor in influencing academic outcomes (i.e., student achievement and motivation) and overall wellbeing in the work environment. Lumpe et al.'s (2011) study on teacher PD programs found that teachers' beliefs and the number of hours spent in PD were predictors of students' achievement in science. Additionally, Bray-Clark and Bates (2003) believed that the PD framework for teachers should emphasize self-efficacy to improve teacher competence to improve student outcomes.

Furthermore, the literature suggests that teachers hold beliefs about their own pedagogical skills or practices that make the integration of CT challenging (Grover & Pea, 2013; Sentence & Csizmadia, 2017; Yadav et al., 2016). Thus, teachers expressed the need for quality teacher development programs to improve and build upon their content knowledge, pedagogy, and confidence to deliver CT (Kong et al., 2020). Hence, PD must build teachers' confidence of their own CT competency and their subsequent ability to foster this thinking in their students. PD must also address teachers' beliefs about their teaching and how it will lead to intended outcomes (i.e., outcome expectancy) (Pajares, 1996; White et al., 2019). These beliefs are influenced by personal experience and observation of role models (Leonard et al., 2018; Schunk, 1991). Both positive self-efficacy and outcome expectancy beliefs shape teachers' confidence in implementing novel teaching approaches, such as CT.

Leonard et al. (2018) conducted the Visualization Basic (i.e., uGame-iCompute) study to examine how teacher beliefs and attitudes toward CT shifted as a result of participating in a study focused on game design and robotics in an afterschool setting. Prior to participating in the study, teachers enrolled in an online graduate course that focused on culturally responsive teaching and either robotics, game design, or blended robotics and game design. The results of the study revealed that teachers' efficacy beliefs and outcomes increased and their CT understanding and dispositions increased. Moreover, several teachers exhibited equitable STEM practices and integrated some elements of their students' culture (e.g., music, Indigenous symbols, etc.) into their afterschool activities. These and other studies suggest that self-efficacy beliefs can positively influence teachers' pedagogical practices relating to CT. Thus, PD should be grounded in helping teachers infuse computer and information science tools into STEM classes and empower them through carefully designed PD that is focused on improving their teaching self-efficacy as well as confidence in their ability to teach the subject matter effectively.

The Bessie Coleman Project

As a part of the Bessie Coleman Project (BCP), elementary and middle school teachers participated in PD experiences that were grounded in culturally responsive teaching practices and CT competencies. Specifically, teachers were led through a series of activities where they flew drones, modeled flight simulations, and created physical models using 3D printing technologies. In these PD sessions, teachers were shown how to integrate these technologies in their out-of-school programs in ways that helped students engage in STEM learning and integrate algorithmic thinking. The teachers were also trained to use the 3D printer, and they encouraged students to create artifacts that reflected their own community and cultural experiences; printing in 3D gave students (and adults) opportunities to make products that are useful in their world.

Research questions that guided the study reported here were as follows: 1. To what extent did teachers' efficacy and STEM attitudes change after participating in the BCP? 2. What pedagogical practices did teachers exhibit during instruction? 3. What suggestions did focal teachers offer to improve PD?

An Equity-Focused Computational Thinking Professional Development Experience

The PD course reflected an equity-focused approach through the integration of culturally specific pedagogy (Leonard, 2019). As a part of this approach, teachers were exposed to various Black pioneers in STEM. Drones were named after Black aviators, astronauts, and NASA personnel, including Bessie Coleman, the first Black and Native-American female pilot and namesake for the project, Katherine Johnson, Mary Jackson, and Dorothy Vaughan. In addition to being trained to fly drones, collect data, and use computational approaches to understand and analyze these data, teachers were trained to use drones to tell a story about their community. For example, one of the teachers shared a drone story as a way to explore her urban community in the eastern United States. Native teachers in Wyoming (one of the project sites) planned a field trip that allowed Arapaho teachers and students to travel across the reservation to a buffalo jump and fly drones in that sacred place. Both of these examples show how drones can be used in culturally specific ways that promote equity for students of color.

Methods

Mixed methods were used to collect and analyze data in the BCP. As it related to teachers and PD, the Teacher Efficacy and Attitudes Toward STEM (T-STEM) Survey for elementary teachers was administered as a pre-post measure to collect quantitative data. The T-STEM Survey consisted of nine constructs: (a) science teaching efficacy and beliefs (STEB); (b) science teaching outcome expectancy (STOE); (c) mathematics teaching efficacy and beliefs (MTEB); (d) mathematics teaching outcome expectancy (MTOE); (e) student technology use (STU); (f) elementary STEM instruction (ESI); (g) 21st-century learning attitudes (CLA); (h) teacher leadership attitudes (TLA); and (i) STEM career awareness (SCA). Each of the subscales was very reliable (Unfried et al., 2014), as Cronbach's alpha ranged from a low of 0.84 to a high of 0.95. Descriptive statistics was used to compare the results of two cohorts of teachers (one in Wyoming and one in Pennsylvania) that participated in the BCP in Year 2. Pilot data for Year 1 were reported elsewhere (Leonard, Jordan et al., 2019).

The DoS tool was also used to rate teachers' lessons as they implemented the program during afterschool clubs in Wyoming and Pennsylvania. The DoS allows observers to collect field notes (qualitative data) and numerical ratings for teachers' performance (quantitative data). The DoS utilizes a 4-point rubric to rate teachers in 12 domains that range from organizational skills to STEM practices, relevance, and relationships (Shah et al., 2014). Scores of 1 or 2 indicate weak practices while scores of 3 or 4 show practices that are appropriate for informal STEM settings (Papazian et al., 2013). Further analysis shows the 12 domains on the DoS can be aggregated into two groups: (a) learning environment (i.e., items 1–3) and (b) student learning (i.e., items 4–12). Additionally, student learning may be further divided into three categories: (a) activity engagement (i.e., items 4–6), (b) STEM knowledge & practices (i.e., items 7–8), and (c) youth development (i.e., items 9–12) (Shah et al., 2014). Transcripts of field notes provided some evidence of teacher performance along with the rating scale.

Finally, teachers participated in focus group interviews after they attended PD in Year 2. However, the structure of the PD in Year 2 was changed based on teacher feedback in Year 1. In the first year, the BCP team had teachers participate in 18 hours of PD over a three-day period that focused on all of the topics that could be taught in the project (i.e., computer modeling, 3D printing, flight simulation, and drones). As a result, teachers concluded that they needed to teach everything in the afterschool program and quickly became overwhelmed, resulting in little content being taught in depth. In Year 2 of the BCP study, the research team focused on fewer topics during the PD, which allowed teachers to gain a deeper understanding of the content. Moreover, teachers of younger students implemented computer modeling and 3D-printing only, while teachers of older students implemented flight simulation and drone navigation only. Furthermore, teachers co-taught in the afterschool program, worked together to connect the content to their students' interest, and spent more time using the technologies.

Results

The results of the BCP are presented in three parts to show (a) impacts of teachers' efficacy and STEM attitudes; (b) teaching strategies to implement innovative technology experiences that enhance students' CT; and (c) how feedback on the PD was used to better prepare future teachers to implement the program more effectively with students. Specifically, the results presented in this chapter are directly related to teachers' experiences in the PD activities and what they did in the afterschool programs as a result of those experiences.

Teacher Self-Efficacy

A total of 12 teachers led afterschool clubs or summer camps during Year 2 of the BCP. Six teachers implemented the project in Philadelphia in spring

TABLE 6.1 Descriptive Data on Pre-Post T-STEM Survey Data

| Construct | Pennsylvania Pretest (n = 6) | | Pennsylvania Posttest (n = 6) | | | Wyoming Pretest (n = 6) | | Wyoming Posttest (n = 6) | | |
	Mean	SD	Mean	SD	Gain	Mean	SD	Mean	SD	Gain
STEB	3.55	0.62	4.26	0.25	0.71	3.77	0.32	3.86	0.41	0.09
STOE	3.74	0.61	3.80	0.53	0.06	3.48	0.43	3.48	0.55	0
MTEB	3.41	0.79	3.92	0.47	0.51	4.06	0.40	4.09	0.47	0.03
MTOE	3.63	0.73	3.74	0.54	0.11	3.81	0.57	3.57	0.60	(0.24)
STU	2.92	1.13	3.67	0.99	0.75	3.00	0.67	2.84	0.82	(0.60)
ESI	2.85	1.07	3.95	0.42	1.10	3.52	0.41	3.31	0.73	(0.21)
CLA	4.64	0.63	4.82	0.40	0.18	4.79	0.20	4.88	0.18	0.09
TLA	4.83	0.41	4.86	0.34	0.03	4.92	0.14	4.00	0.00	(0.92)
SCA	3.75	0.92	4.42	0.47	0.67	3.50	1.04	3.67	1.22	0.17

2019, two teachers facilitated the program in Wyoming in summer 2019, and four teachers implemented the program in Wyoming in fall 2019. All teachers were trained to use Tinkercad and Sculptris to produce 3D artifacts. They also received training to fly drones and to facilitate flight simulation using laptops with joysticks to run the Flight Simulator X program. Project staff administered the T-STEM survey as a pre-post measure. Descriptive statistics are shown in Table 6.1.

The descriptive statistics reveal that teachers in the Philadelphia and Wyoming cohorts had different strengths. However, it should be noted that Pennsylvania teachers received three days (six hours per day) of face-to-face PD led by project staff at one of the schools sites while Wyoming teachers participated in online PD led by project staff during a four-week course (3 hours per week). Although a comparison of scores cannot be attributed to the scope and type of PD alone, the data show several interesting trends. Wyoming teachers had higher STEB, MTEB, MTOE, STU, ESI, CLA, and TLA pretest scores than Pennsylvania teachers. However, Pennsylvania teachers had higher gain scores on these constructs on the posttest. In fact, Pennsylvania teachers had higher gain scores in all categories while Wyoming teachers declined on the constructs of MTOE, STU, CLA, and TLA and had no gain on STOE. Additional data, such as classroom observations and focus group interviews, are needed to explain this result.

Dimensions of Success

The DoS observation tool was used to rate teachers' practices in Year 2. However, most teachers co-taught the students during afterschool or summer school clubs. Feedback was provided to teachers to help them maintain or improve their performance.

TABLE 6.2 BCP Dimensions of Success Ratings (Year 2)

DoS Category	Pennsylvania 1	Pennsylvania 2	Wyoming 1	Wyoming 2	Mean Ratings (STD DEV)
Participation	4	4	4	3	3.75 (0.5)
Purposeful Activities	3	4	4	4	3.75 (0.5)
Engagement with STEM	4	4	4	4	4.00 (0.0)
STEM Content Learning	4	4	3	3	3.50 (0.58)
Inquiry	4	4	4	3	3.75 (0.5)
Reflection	4	2	2	1	2.25 (1.26)
Relationships	4	4	4	3	3.75 (0.5)
Relevance	1	2	4	3	2.50 (1.29)
Youth Voice	4	3	3	4	3.50 (0.58)

Data presented in this chapter are limited to two observations of teachers at each school (i.e., Philadelphia—6 site visits, Wyoming—4 site visits) for a total of ten site visits. Recall that scores of 1 or 2 reflect weak evidence of appropriate STEM practices, and scores of 3 or 4 represent strong evidence of STEM practices. Table 6.2 shows the ratings of four teachers who were selected as a convenience sample. Two DoS reports show ratings for Philadelphia teachers' practices, and two reports reveal the ratings for Wyoming teachers' practices. However, rather than reporting on all 12 DoS categories, we report the nine DoS categories that reflect student learning and engagement in STEM.

Analysis of DoS data reveals teachers' strengths and weakness in terms of STEM practices. Mean ratings across all nine categories show team teachers at both project sites exhibited strong evidence of STEM practices. Further analysis by categories reveal engagement with STEM was the strongest category (M = 4.00 (STD DEV = 0), followed by participation (M = 3.75; STD DEV = 0.5), purposeful activities (M = 3.75; STD DEV = 0.5), inquiry (M = 3.75; STD DEV = 0.5), and relationships (M = 3.75; STD DEV = 0.5). The weakest scores were in the categories of relevance (M = 2.50; STD DEV = 1.29) and reflection (M = 2.25; STD DEV = 1.26). Analysis by teacher for the relevance category revealed two teachers (both in Pennsylvania) neglected to communicate the relevance of the lesson to students' lives. Further analysis by teacher for the reflection category revealed three teachers (one in Pennsylvania and two in Wyoming) did not allow students to have adequate opportunities to reflect on what they were doing during the lessons that were observed. Transcripts of two lessons at each of the project sites (one in Pennsylvania and one in Wyoming) illustrate both strengths and weaknesses during the teaching episodes.

The lesson presented in the first episode occurred at one of three Pennsylvania schools. The school is identified as School C for research purposes. Field notes

FIGURE 6.1 Student Working on a Paper Airplane

were collected in March 2019 as two teachers (Ms. T and Ms. C) co-taught lessons about the principles of flight in the BCP afterschool program with 12 student participants (six boys and six girls), who identified as Black/African American. Students learned about thrust by making paper airplanes that they propelled by blowing into a straw (see Figure 6.1). They also engaged in mathematics by measuring the distance flown in centimeters and recording the best flights.

Episode 1 (Pennsylvania, March 2019)

T: *Today you are going to learn about thrust. You are going to create a plane using straws. Think about what kind of wings are you going to put on it to make it fly straight.*

S1: *How are you supposed to make a paper airplane out of straws?* (female)

T: *We will have to wait and see. We have two different types of straws…big Wawa (i.e., convenience store) and small colored straws. Remember we are talking about thrust. Talk to your group* (i.e., brainstorm ideas).

Students discuss what they could do to make the airplane in small groups of four students at three tables.

T: *How are you going to show thrust? What's thrust?*

S1: *The power.* (female) [The girl shows how she would throw the plane using a forward motion.]

S2: *The way I throw it and how much power I put into it.* (male)

S3: *Ms. T gave me a hint.* (male) [Blows with his mouth.]

S4: *We are going to blow into the straw.* (male)

C: *Step 1—put one straw inside the other and use the tape.*

T: *To make the plane more powerful, designers need to think about how to increase the thrust.*

S5: *You are going to blow into the yellow straw to move the Wawa straw.* (female)

C: *So which one is going to be the airplane?*

S5: *The Wawa straw.* (female)

C: *Get your materials and then we will move to Step 2.*

> Students get materials. They cover the end of the larger straw with tape and practice blowing it out with the smaller straw.

C: *You are going to make a prototype. The expectation is to fly a straight distance and measure it.*

Clearly, the teaching episode given earlier illustrates participation, purposeful activities, STEM content and engagement, inquiry, and strong relationships. Reflection and youth voice were illustrated at the conclusion of the lesson when teachers asked students to recap what they learned. The goal of the BCP was to increase students' knowledge and awareness of STEM and STEM-related careers. Learning about thrust was a direct link to aviation and Bessie Coleman, for whom the project was named. However, at no point during the observation did the teachers explain why the students were learning the activity. Nevertheless, the teaching episode reflects what teachers gained from the PD. They gained content knowledge and the ability to design lessons that related to flight in order to prepare students to successfully engage in the Flight Simulator X program and fly drones. Thus, this lesson provides some insight for understanding Philadelphia teachers' T-STEM gain scores.

Episode 2 (Wyoming, October 2019)

The lesson that was observed in Wyoming was about flight simulation and drones. Two teachers (Ms. J and Mr. W), who had also participated in the pilot study, co-taught the lesson which began with a video and short discussion about drones. The discussion focused on the type of data drones can collect from aerial video: terrain, wildlife, migration of animals, dried up river tributaries, etc. The student participants ($n = 22$; 17 boys and 5 girls) reported on their experiences with flight simulation. Some comments obtained from field notes include the following: *It takes a lot of time for planes to completely stop. [It] took me a long time because angles took a long time to try, like 80°; it was really cool, but it went off the screen; I forgot to do the runway, and I did low and bumps, I did lots of turns so the drone is slower and easier to the land.* Next, the lead teacher, Ms. J, described the activities for the day, which consisted of flight simulation and flying drones. Two additional teachers assisted with the lesson, as students needed to be divided into two groups. A project staff member remained inside to collect field notes, which included student dialogue and interactions that were transcribed as follows:

T: *We want to see if you can follow directions. The weather needs to cooperate with us. This year, we are not using Google Maps; you will fill out a flight plan, there are two tutorials you have to pass before you can fly, and you have missions and more tutorials and more missions.*

[Most students work on Flight Simulator X and four students (all males) go outside to fly drones. Two instructors are inside and two are outside.]

S₁: *It's freezing out there.*

S₂: *Shift, F10! You're not doing it right.*

S₁: *I like the joystick!*

S₂: *Is that how you fly? Are you flying or what?*

S₁: *It's working! I'm going somewhere.*

S₂: *It won't let us move, We're a sitting duck.*

S₁: *You control that, and I'll control this. Control E starts the plane. I got it up in the air finally.*

S₂: *Did you go through the gateways? Don't crash!*

S₁: *Hey, how far did I go?*

T: *Twelve minutes before we start cleaning up.*

S₂: *No, no, no, no, no, crash! You're going to crash.*

[Students are intently working on flight simulation, there is random happy chatter.]

S₃: *Oh no! My engine is stopping.*

S₄: *This plane is so slow, S₃. Dang it!*

S₃: *To get on track you have to go this way.*

[S₃ takes S₄ by the hand and shows him how to do it.]

T: *Log-off computer, plug it in, put simulators in box and flight logs in green basket.*

Students worked rather independently without much teacher guidance in Episode 2. At the outset, teachers focused on relevance as a brief discussion emerged about how drones can be used to collect different types of data (e.g., population density of wildlife; migration patterns; water levels in rivers, etc.). Connections were made to mathematics as some students measured with angles and to science processes as other students maintained flight logs. There was also strong evidence of significant student participation, purposeful activities, and STEM content learning and engagement. The episode revealed that students were highly engaged in the Flight Simulator X program. Generally, students worked together in pairs. From the dialogue, it appears that one student was a bit more critical of the other in one simulation, offering feedback that appeared to be harsh at times. Although the dialogue among the second pair of students was short, there appeared to be collaboration, and one student showed the other how to guide the plane using the joystick. The teaching episode also provides some evidence of inquiry and youth voice. But the simulation ended abruptly without any reflection on the simulation activity and how it connected to program objectives. Thus, it appears that the PD, while online, prepared these teachers to engage students in the Flight Simulator X program and to fly drones. Nevertheless, the online PD may not have provided Wyoming teachers with the computing and leadership skills they needed to impact T-STEM posttest scores.

Focus Group Interviews

Five of the teachers, who participated in the Year 2 study (i.e., completed DoS observations and pre-post surveys), also participated in an earlier PD program during the pilot study in 2018. Unfortunately, no focus group data were collected in Year 2 due to the untimely attrition of the graduate assistant (GA). During the pilot study, teachers were offered a three-day (18 hours) PD course that was face-to-face. The PD was implemented by project staff and three STEM facilitators with expertise in computer modeling, drones, and flight simulation, respectively. The GA administered the interview protocol, which prompted teachers to reflect on the PD they received and to assess its future impact on teaching and learning. The GA also transcribed the focus group interview. Only the reflections of the teachers who participated in Year 2 are presented in this chapter to ensure data integrity.

Teacher Reflection on Professional Development (BCP Pilot Study)

Two teachers, who participated in the Year 2 BCP study, were randomly selected to participate in the focus group interviews. Ms. Davis, who is the subject of the opening vignette and a teacher participant in Pennsylvania, was one of the focal group participants. Mr. Winston (pseudonym) aka Mr. W, who was a teacher participant in Wyoming, was the other. Ms. Davis had 17 years of teaching experience with upper elementary students in urban settings, and Mr. Winston had seven years of kindergarten teaching experience in a rural setting.

The two teachers were asked, "What aspects of this PD have been most helpful to you? In what ways? What are the most important things you've learned from this PD?" They had the following responses:

> One of the things I think in education, in general, that our professional development is boring; to be frank it is boring. We preach that we have to have our students play manipulatives, but then teachers sometimes can be controlling where they tend to think that they have to explain the manipulatives to the kids and show them how to use the manipulatives instead of actually giving the students the manipulatives, having them explore the manipulatives, and helping them kind of evolve and problem solve through what they need to create, and that's not something that I think is as prevalent in professional development as it should be. Since

we had time restrictions this week, if we had two weeks, we would be able to dive deep in to this. I think knowing that you are going to have to learn this with your manipulatives this day; it creates much more fun. The time went so fast and I enjoyed it.

(Ms. Davis)

The level of engagement. This PD has been a lot different due to the hands-on opportunities we have had with drones. Having professionals guiding us throughout the process has been great. Knowing how these resources can apply to real life.

(Mr. Winston)

Indirectly, Ms. Davis implied that the hands-on nature of the PD component of the BCP was a contrast to past PD activities that she received: *I think in education, in general, that our professional development is boring; to be frank it is boring.* She noted that allowing students to create and experience technology should be more prevalent in teachers' PD. She stated: *I think knowing that you are going to have to learn this with your manipulatives this day; it creates much more fun.* Mr. Winston, on the other hand, commented on the role of STEM professionals and noted they were an asset to his PD.

In response to the query "Has the PD been effective in helping you learn more about computer modeling? If so, how?" The two focal teachers had the following responses:

The 3D modeling… being able to get the opportunity to teach that and watching it come alive in front of your screen is amazing. I am still in awe! About knowing this exists and such a way to use it. Learning how to use it is extremely important and being able to discuss with the other teachers how this is best effective in the classroom.

(Ms. Davis)

This PD has been effective in helping me learn computer modeling. The PD goes so fast so personal research and practice will be necessary. When we talk about so many different topics we only have so much time to practice with each resource.

(Mr. Winston)

Both of the foregoing excerpts relate to providing quality learning experiences for computer modeling. Ms. Davis not only mentioned the importance of gaining knowledge but also spoke of how collaboration with other teachers helped her to learn computer modeling for effective classroom instruction. Mr. Winston mentioned how fast the expert went through the instruction and how he compensated by exploring different programming aspects outside of the PD setting.

In reference to drones ("What have you learned about drones that has been helpful to you?"), the responses of the two teachers were quite similar:

> *Yes, when we talked about the drones, we discussed all the regulations that go with the drones, and we also discussed how to make sure you are respectful of people's space, property when you are flying drones, and those are going to be one of the safety pieces that are going to be pertinent for us to teach. As far as computer modeling, I didn't realize that we could do so much on the computer. I don't think I realized until I had my hands on it and explored it.*
>
> (Ms. Davis)

> *Yes. Learning all the background knowledge of flight regulations and parameters to flight. There are certain steps to take to go through this process the correct way.*
>
> (Mr. Winston)

Both teachers focused on regulations and safety when they reflected on drones. Thus, the PD was successful in terms of increasing teachers' awareness of safety standards for drone use and ensuring drones were used appropriately. As evident in Episode 2, students were supervised during drone flight and only small groups of students were allowed to fly the drones at any given time.

Focal teachers were also asked, "How has the PD been relevant to your classroom instruction? What aspects of the PD do you foresee potentially incorporating into your instruction?" Ms. Davis and Mr. Winston responded as follows:

> *So, when I get to commence in the classroom, I am going to use this in the school district. I am going to show this not only to my classroom but show this to the whole school. Introduce this to the whole school. I am also going to introduce to the middle school and high school. I am going to utilize this to encourage the academic growth with our school district.*
>
> (Ms. Davis)

> *Fourth and fifth graders are going to have a blast working with 3D printers and having the opportunity to fly drones. 3D printing will be the most applicable. Math and scaling can be applied to a real-world application. Creating high engagement for the students is so exciting, and they will have a blast with the drones.*
>
> (Mr. Winston)

Understanding teachers' perception of the applicability of the PD to classroom instruction was of importance to project staff. In the given excerpts, Mr. Winston demonstrated that he was able to make connections between project activities and the teaching and learning of academic subjects. Specifically, Mr. Winston envisioned ways to connect mathematics content to the technology applications. This was evident in Teaching Episode 2 when students used angles and flight plans to document flight simulation. Students were able to use CT as they tinkered with the flight simulation program and learned through trial-and-error.

During interviews, staff surveyed the teachers about the degree to which the sessions were effective in helping them to learn the theoretical importance of culturally responsive pedagogy and how best to infuse new knowledge and ideas into their lessons. When asked how the PD helped to bolster her understanding of culturally responsive pedagogy, Ms. Davis, who was the only focal teacher from Philadelphia to participate in Year 2, responded to CRP as follows:

> *I would say kids can do everything in my classroom. We always say first, everything is possible and there aren't any limits. When everything is possible and there aren't any limits, we are going to be flying drones at the end of the school year. Along with flying drones at the end of the school year, they are also going to know their coordinate geometry, they are also going to learn about probability, they are also going to be able to understand number and concepts and their conversion measurements. I am that excited, I cannot wait to get back to write a lesson plan about this. The geography [and connections to] wind speed, time, and distance for a 10-year-old is amazing.*

Ms. Davis's response as contextualized here was also noted in the opening vignette. Her beliefs about students of color and their capacity to learn were connected to her ideology about culturally responsive pedagogy. Ms. Davis believed drone applications would improve students' mathematics content knowledge because there were no limits to what her students could do. Going a step further to use drone footage to inform and make children's communities better, drone applications would also lead to computational participation.

Finally, although the PD took place on a college campus for intensive training over a three-day period, comments about the overall quality of the PD followed a consistent theme related to more time:

> *I think if the PD last longer that's the only thing, I do understand the finances and everything else, but I would love to stay for two weeks to work on this, because this is introductory. Positively, I would say I am honored to be chosen for this; I want more of this. I want to be able to take more of this to more universities. I love my school. I know that we would take this into great heights. The school district I work for in [Pennsylvania] should be in the forefront of this because we had an opportunity to [participate] in this.*
>
> (Ms. Davis)

> *Giving more time to practice with computer modeling and have step-by-step instruction really helps us learn the process. Building a strong foundation for us will allow us to be better teachers to our students.*
>
> (Mr. Winston)

Although there were some time constraints associated with the PD that took place in the BCP, these teachers' request for additional training should not go unheeded. Future research endeavors should examine the impact of moderate (12–18 hours) versus substantive (24–36 hours) and online versus face-to-face PD.

Discussion

In this chapter, we provided information on how to design and implement PD experiences that help teachers to develop and understand CT/participation and dispositions in ways that foster these competencies in students of color. Then we highlighted results from the BCP, which provided underrepresented students with unique opportunities to participate in flight simulation and flying drones to increase their awareness of and interest in STEM careers. A critical component of the project was providing teachers with culturally responsive PD experiences that allowed them to successfully implement the project.

The results of this study reveal three important findings as they relate to PD in the BCP. These findings emerged from the quantitative and qualitative data presented earlier. The first finding relates to the T-STEM survey. Results revealed there were gains on all nine constructs for six Philadelphia teachers. The highest gains were made in elementary science instruction (ESI), student technology use (STU), and science teaching efficacy beliefs (STEB). For the Wyoming teachers, there were gains on four constructs: STEB, mathematics teaching efficacy beliefs (MTEB), 21st-century learning attitudes (CLA), and STEM career awareness (SCA). However, there were declines on four constructs: mathematics teaching outcome expectance (MTOE), STU, and ESI. Scores on the construct of science teaching outcome expectancy (STOE) were unchanged for Wyoming teachers. Although these results are based on a small data set ($n = 12$), supplemental DoS observations and focus group interviews suggest that the online PD and the shortened time frame may have influenced the decline on some of the Wyoming teachers' T-STEM scores. Overall, the findings are promising, especially for teachers who participated in the ITEST studies (i.e., uGame-iCompute and BCP) more than once.

The T-STEM results confirmed that engaging teachers in out-of-school PD that focused on CT/participation has positive impacts on STEB and MTEB beliefs. Gains in these two constructs occurred for both the Philadelphia and Wyoming teachers. This finding, which is consistent with prior studies, implies that the PD and out-of-school experiences improved teachers' beliefs about their ability to effectively teach activities related to mathematics and science, as well as their beliefs about how effective they could be in impacting student learning in STEM (Leonard, Chamberlin et al., 2019). Specifically, allowing teachers to engage in hands-on activities where they learned how to use the technology in ways that enriched their mathematics understanding (i.e., 3D printing, scaling, etc.) and science content knowledge (i.e., principles of flight, air pressure, weather, etc.) had an impact on teachers' STEB and MTEB posttest scores (Leonard, Chamberlin et al., 2019; White et al., 2019).

Most importantly, infusing culturally responsive pedagogy with STEM had a positive impact on teachers' sense of STEM career awareness as well as their understanding of ways to foster this interest in their students. One of the major elements of the PD was to expose teachers to Black pioneers in STEM (i.e., culturally responsive pedagogy), not just through naming the drones after these noteworthy figures but also by connecting their stories to the content and technologies used in the project. In addition, the PD included discussions and explorations of how drones, flight simulation, and 3D printing were used in various careers. Furthermore, it provided teachers with strategies to help students to connect these technologies to careers in STEM.

The second finding relates to the DoS observation tool. Results, which were similar to previous findings (Leonard et al., 2018), revealed mean DoS scores indicated appropriate STEM instruction in eight of the nine student learning categories. Students exhibited evidence of STEM engagement on all constructs except relevance. Teachers in both cohorts scored the highest on STEM engagement, participation, and purposeful activities. The excerpts from the lessons illustrate how teachers were able to use hands-on activities to teach flight-related concepts. These activities were extensions of what teachers learned in the BCP PD workshops. Both the DoS ratings and field notes show how teachers were able to translate their own learning from the PD into meaningful and responsive learning experiences for their students that were hands-on and purposeful. The experiences also fostered active participation to build understanding of important STEM content (e.g., thrust, data collection, and data analysis).

Additionally, there was evidence from both the DoS and the field notes that teachers were able to facilitate computational participation and dispositions among the students (Kafai & Burke, 2014; Pérez, 2018). In the PD, teachers engaged in hands-on experiences where they had to work together to not only learn the technologies but also to develop lesson plans and strategies that would support student learning and engagement in the project. The type of participation that was experienced in the PD was fostered in the out-of-school activities that were implemented by the teachers and demonstrated by the students as they worked together to solve content focused and technical problems. The students also demonstrated positive dispositions by their willingness to persist through challenges and collaborate with others to achieve a goal (DiCerbo, 2016; Pérez, 2018).

The third finding draws from the focus group interviews to provide significant insight into key features of PD that can support teachers in their integration of CT into their instructional practices. First, given that the PD experience was a part of an out-of-school project, the teachers were not constrained by the parameters of the school curricula. This kind of PD, although intended for informal learning, allowed Ms. Davis and Mr. Winston to make various connections to the curricula in creative ways. Both Ms. Davis and Mr. Winston commented that, generally, PD is boring or rigid, and provides few opportunities for hands-on activities and active engagement. In contrast, they found the PD provided through the BCP to be fun and engaging, and they noted that it provided opportunities for hands-on learning through the use of emerging technologies. Both teachers acknowledged that they wished they had more time to engage in the PD. Yet, they were grateful for the opportunity to participate and saw how this work would enhance their teaching practices.

Lastly, the PD experience provided an opportunity to denounce deficit narratives of rural students and students of color and to affirm positive narratives about students of color (Delpit, 2012). Both Ms. Davis and Mr. Winston saw themselves as advocates for their students and actively challenged notions that their students were deficient in any way. Further, they believed that their students could demonstrate proficiency in mathematics and thought of ways to utilize the content they were using in the afterschool program to develop challenging mathematical topics. Because CT and culturally responsive pedagogy were integrated into the PD, the teachers were able to reflect on the content they were asked to teach as well as how to teach it in ways that were engaging and culturally responsive to their students. Just as Ms. Davis and Mr. Winston saw their students as capable and limitless in their ability to rise to challenges, it is important that all teachers of students of color take a similar stance to encourage persistence in problem solving and engagement in critical thinking skills to help their students develop CT (Kafai & Burke, 2014, 2017). PD that integrates culturally responsive pedagogy with real-world experiences, innovative technologies, and CT competencies allows teachers to deepen social justice perspectives and critical consciousness that will lead to better outcomes for all students.

Summary

In this chapter, we highlighted PD literature that supports teachers by developing competencies in CT/participation and dispositions, building on teacher's knowledge and experiences, promoting positive self-efficacy, and grounding PD in culturally responsive pedagogy and equity-focused narratives of students of color. As cited in the literature, PD that promotes teachers' use of CT must be grounded in culturally responsive, equity-centered perspectives that focus on asset-based narratives of children of color (Bower et al., 2017; Darling-Hammond et al., 2019; Joseph et al., 2015). Professional development experiences should focus not only on teacher knowledge of and competencies with CT but also on knowledge and competencies that are grounded in their own experiences, their understanding of students' experiences, and equity-focused, culturally responsive perspectives (Bhaduri et al., 2018; Buss & Gamboa, 2017; Leonard et al., 2016). To do this, PD must help teachers cultivate these competencies themselves, critically reflect on their experiences in ways that foster a willingness to experiment with new practices, and make meaningful and sustained changes in their practices that produce positive outcomes for their students (Clark & Hollingsworth, 2002).

The approach to PD forwarded in this chapter integrated what we already knew about effective practices. We applied those practices to support teachers in ways that promoted CT/participation and dispositions in themselves and, subsequently, in their students. This moves us beyond the traditional forms of PD that relies on learning content and grounds our work in learning real-world applications and emerging technologies that focus on creative ways for teachers to apply what is learned to their instructional practices. Such applications and technologies also foster the kind of collaboration that builds a professional community among teachers that can be translated into creating similar communities among their students. As we move from CT as individuals to computational participation as a community of learners, it is important that teachers are empowered to learn how to use and manage the technologies and apply them purposefully to learn about the communities themselves and offer solutions that speak to community-based issues. Opportunities for impact include neighborhood redevelopment, climate change, and so on. Such projects, while socio-political in nature, have the potential to improve teachers' PD experiences and be effectively translated into their instructional practices. The BCP project illustrates how out-of-school programs can be fertile ground for these kinds of PD opportunities by pushing the limits on what teachers and students can do.

In conclusion, by focusing on teachers' informal learning, building their self-efficacy, helping them reshape narratives about underrepresented students (i.e., female, rural, students of color), and engaging them in real-world activities that foster their own CT/participation and disposition (Desimone & Garet, 2015), we can better prepare and empower them to realize the goal of building computational literacy for all students.

References

Aho, A. V. (2012). Computation and computational thinking. *The Computer Journal*, *55*(7), 832–835.

Asghar, A., Ellington, R., Rice, E., Johnson, F., & Prime, G. M. (2012). Supporting STEM education in secondary science contexts. *Interdisciplinary Journal of Problem-Based Learning*, *6*(2), 85–125.

Avery, Z. K., & Reeve, E. M. (2013). Developing effective STEM professional development programs. *Journal of Technology Education*, *25*(1), 55–69.

Bandura, A. 1997. *Self-efficacy: The exercise of control*. New York: Freeman.

Barni, D., Danioni, F., & Benevene, P. (2019). Teachers' self-efficacy: The role of personal values and motivations for teaching. *Frontiers in Psychology*, *10*(1645). doi:10.3389/fpsyg.2019.01645.

Battey, D. (2013). Access to mathematics: "A possessive investment in whiteness". *Curriculum Inquiry*, *43*(3), 332–359.

Battey, D., & Franke, M. (2015). Integrating professional development on mathematics and equity: Countering deficit views of students of color. *Education and Urban Society*, *47*(4), 433–462.

Bhaduri, S., Gendreau, A., Koushik, V. S., Sumner, T., Ristvey, J., & Russell, R. (2018). Promoting middle school students' motivation and persistence in an after-school engineering program. In J. Barnes-Johnson & J. Johnson (Eds.), *STEM 21: Equity in teaching and learning to meet global challenges of standards, engagement and transformation* (pp. 138–162). Peter Lang.

Bower, M., Wood, L. N., Lai, J. W., Howe, C., Lister, R., Mason, R., Highfield, K., & Veal, J. (2017). Improving the computational thinking pedagogical capabilities of schoolteachers. *Australian Journal of Teacher Education*, *42*(3), 53–72.

Bray-Clark, N., & Bates, R. (2003). Self-efficacy beliefs and teacher effectiveness: Implications for professional development. *The Professional Educator*, *26*(1), 13–22.

Brennan, K., & Resnick, M. (2012, April). New frameworks for studying and assessing the development of computational thinking. In *Proceedings of the 2012 Annual Meeting of the American Educational Research Association, Vancouver, Canada*, (Vol. 1, p. 25).

Buss, A., & Gamboa, R. (2017). Teacher transformations in developing computational thinking: Gaming and robotics use in after-school settings. In P. J. Rich & C. B. Hodges (Eds.), *Emerging research, practice, and policy on computational thinking* (pp. 189–203). Springer.

Celedón-Pattichis, S., Borden, L. L., Pape, S. J., Clements, D. H., Peters, S. A., Males, J. R., Olive, C., & Leonard, J. (2018). Asset-based approaches to equitable mathematics education research and practice. *Journal for Research in Mathematics Education*, *49*(4), 373–389.

Clark, D., & Hollingsworth, H. (2002). Elaborating a model of teacher professional growth. *Teaching and Teacher Education*, *18*(8), 947–967.

Clark, L. M., Frank, T. J., & Davis, J. (2013). Conceptualizing the African American mathematics teacher as a key figure in the African American education historical narrative. *Teachers College Record*, *115*(2), 1–29.

Darling-Hammond, L., Flook, L. F., Cook-Harvey, C., Barron, B., & Osher, D. (2019). Implications for educational practice of the science of learning and development. *Applied Developmental Science*, *24*(2), 97–140.

Delpit. L. (2012). *"Multiplication is for White People": Raising expectations for other people's children*. The New Press.

Desimone, L. M., & Garet, M. S. (2015). Best practices in teacher's professional development in the United States. *Psychology, Society, & Education*, *7*(3), 252–263. https://pdfs.semanticscholar.org/31ff/d06b4df5bb399f782d3985f17311d2bc44ae.pdf

DiCerbo, K. E. (2016). Assessment of task persistence. In Y. Rosen, S. Ferrar, & M. Mosharraf (Eds.), *Handbook of research on technology tools for real-world skill development* (Vol. 2, pp. 778–804). IGI Global.

Dietz, G., Landay, J. A., & Gweon, H. (2019). *Building blocks of computational thinking: Young children's developing capacities for problem decomposition* [Paper presentation]. *41st Annual Meeting of the Cognitive Science Society, Montreal, Quebec*. https://sll.stanford.edu/docs/2019_cogsci/Dietz_Landay_Gweon_CogSci2019.pdf

Ellington, R. (2019). Toward a transformative framework for STEM education: Achieving equity through a holistic approach. In G. Prime (Ed.), *Effective STEM education for African-American K–12 learners* (pp. 35–70). Peter Lang.

Foster, I. (2006). A two-way street to science's future. *Nature*, *440*(7083), 419–419.

García-Peñalvo, F. J., Reimann, D., Tuul, M., Rees, A., & Jormanainen, I. (2016, October 6). *An overview of the most relevant literature on coding and computational thinking with emphasis on the relevant issues for teachers* [Research report]. TACCLE3 Consortium, Brussels, Belgium. doi:10.5281/zenodo.165123.

Gay, G., & Kirkland, K. (2003). Developing cultural critical consciousness and self-reflection in preservice teacher education. *Theory into Practice, 42*(3), 181–187.

Gholson, M. L. (2017). The mathematical lives Black children: A sociocultural-historical rendering of Black brilliance. In J. Leonard & D. B. Martin (Eds.), *The brilliance of Black children in mathematics: Beyond the numbers and toward new discourse* (pp. 55–76). Information Age Publishing.

Grover, S., & Pea, R. (2013). Computational thinking in K–12: A review of the state of the field. *Educational Researcher, 42*(1), 38–43.

hooks, b. (1994). *Teaching to transgress: Education as the practice of freedom.* Routledge.

Hunsaker, E. (2020). Computational thinking. In A. Ottenbreit-Leftwich & R. Kimmons (Eds.), *The K–12 educational technology handbook.* EdTech Books. https://edtechbooks.org/k12handbook/computational_thinking

Jacobs, V. R., Franke, M. L., Carpenter, T. P., Levi, L., & Battey, D. (2007). Professional development focused on children's algebraic reasoning in elementary school. *Journal for Research in Mathematics Education, 38*(3), 258–288.

Joseph, N. M., Haynes, C. M., & Cobb, F. (Eds.). (2015). *Interrogating whiteness and relinquishing power: White faculty's commitment to racial consciousness in STEM classrooms.* Peter Lang.

Kafai, Y. B. (2016). From computational thinking to computational participation in K–12 education. *Communications of the ACM, 59*(8), 26–27.

Kafai, Y. B., & Burke, Q. (2014). *Connected code: Why children need to learn programming.* The MIT Press.

Kafai, Y. B., & Burke, Q. (2017). Computational participation: Teaching kids to create and connect through code. In P. J. Rich & C. B. Hodges (Eds.), *Emerging research, practice, and policy on computational thinking* (pp. 393–405). Springer.

Kafura, D., Bart, A., & Chowdhury, B. (2015, June). Design and preliminary results from a computational thinking course. In *Proceedings of the 2015 ACM Conference on Innovation and Technology in Computer Science Education* (pp. 63–68). https://doi.org/10.1145/2729094.2742593

Ketelhut, D. J., Mills, K., Hestness, E., Cabrera, L., Plane, J., & McGinnis, J. (2019). Teacher change following a professional development experience in integrating computational thinking into elementary science. *Journal of Science Education and Technology, 29*(1), 174–188. doi:10.1007/s10956-019-09798-4

Killen, H., Coenraad, M., Byrne, V., Cabrera, L., Ketelhut, D. J., & Plane, J. (2020, June 19–23). *Reimagining computational thinking professional development: Benefits of a community of practice model* [Paper presentation]. International Society for the Learning Sciences, Nashville, TN, United States. https://repository.isls.org/bitstream/1/6502/1/2125-2132.pdf

Kong, S., Lai, M., & Sun, D. (2020). Teacher development in computational thinking: Design and learning outcomes of programming concepts, practices and pedagogy. *Computers & Education, 151*(1), 1–19.

Lee, I., Martin, F., & Apone, K. (2014). Integrating computational thinking across the K–8 curriculum. *ACM Inroads, 5*(4), 64–71.

Leonard, J. (2019). *Culturally specific pedagogy in the mathematics classroom: Strategies for teachers and students.* Routledge.

Leonard, J., Buss, A., Gamboa, R., Mitchell, M., Fashola, O. S., Hubert, T., & Almughyirah, S. (2016). Using robotics and game design to enhance children's self-efficacy, STEM attitudes, and computational thinking skills. *Journal of Science Education and Technology*, *25*(6), 860–876.

Leonard, J., Chamberlin, S., Aryana, S., Even, A., & Lazic, M. (2019). Using STEM internships to recruit and retain Noyce scholars in elementary education. In J. Leonard, A. C. Burrows, & R. Kitchen (Eds.), *Recruiting, preparing, and retaining STEM teachers for a global generation* (pp. 3–35). Brill/Sense.

Leonard, J., Chamberlin, S., Bailey, B. E., Verma, G., & Douglass, H. (2019). Broadening millennials' participation in STEM and teaching professions through culturally relevant, place-based, informal science internships. In G. Prime (Ed.), *Effective STEM education for African-American K–12 learners* (pp. 95–128). Peter Lang.

Leonard, J., Jordan, W. J., & Ellington, R. (2019, November). The Bessie Coleman Project: Using computer modeling and flight simulation in informal STEM settings. In S. Otten, A. G. Candela, Z. de Araujo, C. Haines, & C. Munter (Eds.), *Proceedings of the 41st Annual Meeting of the North American Chapter of the International Group for the Psychology of Mathematics Education* (pp. 106–110). University of Missouri.

Leonard, J., Mitchell, M., Barnes-Johnson, J., Unertl, A., Outka-Hill, J., Robinson, R., & Hester-Croff, C. (2018). Preparing teachers to engage rural students in computational thinking through robotics, game design, and culturally responsive teaching. *Journal of Teacher Education, 69*(4), 386–407.

Li, Q., Richman, L., Haines, S., & McNary, S. (2019). Computational thinking in classrooms: A study of a PD for STEM teachers in high needs schools. *Canadian Journal of Learning and Technology/La revue Canadienne de l'apprentissage et de la technologie, 45*(3), 1–21.

Lumpe, A. T., Czerniak, C., Haney, J., & Beltyukova, S. (2011). Beliefs about teaching science: The relationship between elementary teachers' participation in professional development and student achievement. *International Journal of Science Education, 34*(2), 153–166.

Lye, S. Y., & Koh, J. H. L. (2014). Review on teaching and learning of computational thinking through programming: What is next for K-12? *Computers in Human Behavior, 41*, 51–61.

Martin, D. B. (2009). Researching race in mathematics education. *Teachers College Record, 111*(2), 295–338.

Martin, D. B. (2012). Learning mathematics while Black. *Educational Foundations, 26*(1–2), 47–66.

Mitchell, A. (1998). African American teachers: Unique roles and universal lessons. *Education and Urban Society, 31*, 104–122.

National Research Council. (2011). *Committee for the workshops on computational thinking: Report of a workshop on the pedagogical aspects of computational thinking*. National Academies Press.

Pajares, F. (1996). Self-efficacy beliefs in academic settings. *Review of Educational Research, 66*(4), 543–578.

Papazian, A. E., Noam, G. G., Shah, A. M., & Rufo-McCormick, C. (2013). The quest for quality in afterschool science: The development and application of a new tool. *Afterschool Matters, 18*, 17–24.

Pérez, A. (2018). A framework for computational thinking dispositions in mathematics education. *Journal for Research in Mathematics Education, 49*(4), 424–461.

Schunk, D. H. (1991). Self-efficacy and academic motivation. *Educational Psychologist*, *26*(3–4), 207–231.

Schunk, D. H. (2020). *Learning theories: An educational perspective* (8th ed.). Pearson.

Selby, C. C., & Woollard, J. (2013, July 1–3). *Computational thinking: The developing definition* [Paper presentation]. Innovation and Technology in Computer Science Education Conference, Canterbury, England.

Sentence, S., & Csizmadia, A. (2017). Computing in the curriculum: Challenges and strategies from a teacher's perspective. *Education and Information Technologies*, *22*(2), 469–495.

Shah, A. M., Wylie, C. E., Gitomer, D., & Noam, G. (2014). *Development of the dimensions of success (DoS) observation tool for the out of school time STEM field: Refinement, field-testing and establishment of psychometric properties.* Partners in Education and Resilience. http://www.pearweb.org/research/pdfs/DoSTechReport_092314_final.pdf

Siwatu, K. O. (2007). Preservice teachers' culturally responsive teaching self-efficacy and outcome expectancy beliefs. *Teaching and Teacher Education*, *23*(7), 1086–1101.

Tran, Y. (2018). Computational thinking Equity in elementary classrooms: What third-grade students know and can do. *Journal of Educational Computing Research*, *57*(1), 3–31.

Unfried, A., Faber, M., Townsend, L., & Corn, J. (2014). *Validated student, teacher, and principal survey instructions for STEM education programs* [Conference paper]. American Educational Research Association Annual Meeting, Philadelphia, PA, United States. https://comm.eval.org/HigherLogic/System/DownloadDocumentFile. ashx?DocumentFileKey=0fa6c419-e181-46a4-9e7e-75beaabf0416&forceDialog=0

Wager, A. A., & Foote, M. Q. (2013). Locating praxis for equity in mathematics: Lessons from and for professional development. *Journal of Teacher Education*, *64*(1), 22–34.

Weintrop, D., Beheshti, E., Horn, M., Orton, K., Jona, K., Trouille, L., & Wilensky, U. (2016). Defining computational thinking for mathematics and science classrooms. *Journal of Science Education and Technology*, *25*(1), 127–147.

White, D. Y., Leonard, J., Chamberlin, M., & Buss, A. (2019). Supporting Noyce scholars' teaching of mathematics in rural elementary schools. In J. Leonard, A. C. Burrows, & R. Kitchen (Eds.), *Recruiting, preparing, and retaining STEM teachers for a global generation* (pp. 133–162). Brill/Sense.

Wing, J. (2008). Computational thinking and thinking about computing. *Philosophical Transactions of the Royal Society A: Mathematical, Physical and Engineering Sciences*, *366*(1881), 3717–3725. http://dx.doi.org/10.1098/rsta.2008.0118

Wing, J. M. (2006). Computational thinking. *Communications of the ACM*, *49*(3), 33–35. doi:10.1145/1118178.1118215

Yadav, A., Gretter, S., Good, J., & McLean, T. (2017). Computational thinking in teacher education. In P. J. Rich & C. B. Hodges (Eds.), *Emerging research, practice, and policy on computational thinking* (pp. 205–220). Springer.

Yadav, A., Gretter, S., Hambrusch, S., & Sands, P. (2016). Expanding computer science education in schools: Understanding teacher experiences and challenges. *Computer Science Education*, *26*(4), 235–254.

Yadav, A., Hong, H., & Stephenson, C. (2016). Computational thinking for all: pedagogical approaches to embedding 21st century problem solving in K–12 classrooms. *TechTrends*, *60*(6), 565–568.

Yang, H. (2019). The effects of professional development experience on teacher self-efficacy: Analysis of an international dataset using Bayesian multilevel models. *Professional Development in Education*, *46*(8), 1–15. doi:10.1080/19415257.2019.1643393

7

PROGRAM EVALUATION OF BROADENING STEM PARTICIPATION FOR UNDERREPRESENTED STUDENTS OF COLOR

Monica B. Mitchell and Olatokunbo S. Fashola

Similar to most students attending community college, Clara understands the importance of education as a vehicle for social mobility and derives personal fulfillment from academic success. She had always gravitated toward science and mathematics in school and had been among the few Black students enrolled in advanced placement courses at her high school. Upon graduating from high school, she took a break from college to work and travel for a few years. Ready to start college, Clara relocated to the west coast and enrolled at the local community college in a large urban city. She was confident the local community college would be a good fit. The student population was extremely diverse, and she felt connected to the institution since her great uncle of Mexican-American heritage had studied there in the 1960s. As a Black student of multiethnic ancestry, Clara knew she would feel at home and do well. She enrolled in the chemistry program at the community college and intended to transfer to a four-year institution for her Bachelor's degree once she completed her Associate's degree. While she was not exactly sure of her specific career path, she knew it had to involve science.

Among Clara's first classes at the community college was calculus, a course requirement for all chemistry majors. Having taken four years of college preparatory mathematics in high school, Clara felt she had the fundamental skills under her belt to do well. She jumped right into the course, enjoyed working the solutions to problems, and spent much time and effort completing the assignments and submitting them on time. She studied hard for the midterm. Afterwards, she felt good about her performance on the test. But when the results came back, she was stunned. She had not passed. There had to be a mistake! Clara reviewed her work against the answer key and could not believe her eyes. Her answers were correct, but she had received a zero on almost all of her work.

Soon other students were emailing and texting each other asking how their tests were graded. Students began complaining to one another that they, too, had arrived at the correct answers according to the answer key but had received zeros on several of the problems. Clara knew this was not fair. She had shown all work for how she arrived at the solution to every problem on the midterm. Her answers matched those on the answer key. Therefore, she reached out to the professor, hoping there had to be some mistake. Instead, she was told, "You didn't work the problems the way you were supposed to. It does not matter that you got the right answer or even how you got the right answer. If you do not work the problems exactly as I do, you will get a zero every time." The professor went on to say, "How dare you question my integrity after three decades of teaching students at this community college."

Clara wanted to have a chance to explain her thinking on the problems and why she chose the strategies she did in order to arrive at the solutions. If she got the correct answer, she knew her approach had to have some merit beyond a fat zero. Certainly, the professor would consider assigning partial credit to recognize arriving at the correct answer by using an alternate strategy. According to the professor, partial credit was simply not an option. The professor refused to change any of her scores on the midterm. Exasperated, Clara felt defeated. She certainly did not want to fail calculus, but this professor seemed completely unreasonable. Instead of trying to help students succeed, Clara felt the professor's grading approach was a set-up for failure. Maybe she would have a better experience with another professor, so Clara withdrew from the calculus course.

Mathematics is featured as the subject matter content in the vignette for several reasons. First, in formal schooling, mathematics represents a foundational and common experience among all students. Second, mathematics, as a subject area, continues to serve as a gatekeeper in science, mathematics, technology, and engineering (STEM) that limits the participation of students of color in scientific and technological pathways (Martin et al., 2010; Oakes et al., 2003; Oakes et al., 1990; Stinson, 2004). The vignette demonstrates such gatekeeping practices. Third, while Clara's encounter in introductory calculus as an Afro-Latina seemed draconian and archaic, the vignette depicted an actual experience that occurred during the 2019–2020 academic year.

The vignette also has particular resonance with the two co-authors of this text, Monica and Toks, whose journeys from female students of color in STEM to project evaluators aligned with Clara's experience as a mathematics student. Their experiences, which are not dissimilar to those of underrepresented minority students in STEM, have often been characterized by implicit and explicit deficit-oriented practices, which serve to marginalize students of color and stifle their success (DeGruy, 2005; Gutiérrez & Rogoff, 2003; Martin et al., 2010). With terminal degrees in quantitative-dependent fields, Monica and Toks experienced non-affirming dynamics in the teaching and learning of mathematics as Black girls and women interested in pursuing

STEM. Consequently, their work in STEM evaluation seeks to examine and shed light on educational efforts and interventions designed to positively change the narrative of STEM teaching and learning for students of color. Evaluators of color, who are members of underrepresented communities and have journeyed with and in those communities to advance equity and social justice in STEM, possess a unique and specialized base of knowledge and experience to draw upon.

Recent advances for the adoption of computing and the integration of computational thinking into PreK–12 education brings both opportunities and challenges for evaluating the process and results of these reforms. Computational thinking (CT) is closely related to mathematics (National Research Council, 2010). CT shares commonality with mathematics through components that foster "habits of mind" and problem solving. This relationship is not overlooked in Wing's characterization of CT when she states: "Computer science inherently draws on mathematical thinking, given that, like all sciences, its formal foundations rest on mathematics" (2006, p. 35). According to Wing (2006), "computational thinking is using heuristic reasoning to discover a solution" (p. 34). Heuristic reasoning has roots in Pólya's influential *How to Solve It* (1945), a publication which holds important significance in mathematics education and pedagogy (Leinhardt & Schwarz, 1997; Schoenfeld, 2018). Some suggest a role may exist for CT in broadening the participation of underrepresented minorities in STEM and STEM-related pathways (Bobb & Brown, 2017; Margolis et al., 2014; Pearson, 2009).

The term broadening participation refers to rectifying disproportionate patterns of representation by engaging persons who have been marginalized in STEM—specifically females and underrepresented minorities. The redress of underrepresentation must occur along the entire spectrum of educational and professional pathways, otherwise the promise of broadening participation will remain elusive. The next section addresses broadening STEM participation in more detail.

Evaluation and Broadening STEM Participation

The scientific and technological domain in the U.S. continues to suffer from a lack of diversity and inclusion. The underutilization of diverse talent is seen in persistent disproportionate participation rates of racial and ethnic minorities in the science and engineering workforce (National Science Board [NSB], 2018). In keeping with the equity and equality demands of a democratic society, broadening participation involves the pursuit of the full participation of individuals of color from underserved communities (i.e., Blacks/ African Americans, Latinx/Hispanics, American Indians, Alaska Natives, Native Hawaiians, and other Pacific Islanders) in concordance with their proportional representation in the U.S. population (Committee on Equal Opportunities in

Science and Engineering [CEOSE], 2014; Garrison, 2013; Institute of Medicine, 2010). Importantly, broadening participation also refers to increasing the STEM participation of females and persons with disabilities (Bellman et al., 2018; Dasgupta & Stout, 2014), as well as institutions serving large populations of underrepresented students disadvantaged in the STEM enterprise by lack of resources and/or structural inequities (e.g., underserved school districts) (CEOSE, 2019).

The purpose of this chapter is to share research and evaluation methods that are appropriate for broadening the participation of racial and ethnic minority groups that are historically underrepresented in STEM fields and careers. The latest national data show that among those employed full-time in the 2017 science and engineering (S&E) workforce,[1] the participation of racial and ethnic minorities continued to lag behind their representation in the U.S. population (National Science Foundation [NSF], 2019). For example, the demographics of the full-time S&E workforce in 2017 was 8.6% Latinx/Hispanic, 7.7% Black/African American, 0.3% American Indian or Alaska Native, and 0.3% Native Hawaiian or Other Pacific Islander. Thus, participation rates lagged behind their respective 17.6%, 13.9%, 1.7%, and 0.4% representation in the U.S. population during the same period (NSF, 2019; U.S. Census Bureau, 2017).

Considered an agency priority of the National Science Foundation (NSF), broadening participation is included in its merit review process as an element of the broader impacts criterion.[2] While the use of the term as part of the merit review process implies an overarching or inclusionary construct, the application of broadening participation extends to all aspects of STEM inclusive of PreK-12 schooling, undergraduate education, informal learning, faculty development, and scientific research (Gonzalez, 2014). The establishment of the Committee on Equal Opportunities in Science and Engineering (CEOSE) by Congress in 1980 provided much of the guidance for NSF to address broadening participation.[3]

Program evaluation has an important role to play in broadening STEM participation (Mertens & Hopson, 2006). As defined by Thomas and Campbell (2021), evaluation is "disciplined inquiry involving the systematic, contextually responsive, and ethical application of research tools and methods to collect data that assess the effectiveness and operations of programs within the various social, political, and cultural contexts in which they operate" (p. 6). While sharing similar methods of systematic disciplined inquiry, evaluation and applied social science research are not necessarily the same. The primary purpose of evaluation examines the value or merit of a program or intervention (Mathison, 2008; Scriven, 1991). Typically, program evaluation involves formative (process and progress) and summative (outcomes and results) components of interest to a range of stakeholders, including implementers, funders, participants, organizations, school systems, and communities. Interdisciplinary in nature, program evaluation can be complex and multifaceted. The lifespan of an evaluation entails multiple phases, which include a planning stage, design and conceptualization, implementation, analysis, reporting and use (Goodyear et al., 2019). Program evaluators often interact closely with principal investigators and project directors to align evaluation design with the intended theory of change and to ensure responsiveness of the evaluation as the realities of project implementation unfold.

The computer science education movement represents a more recent but vital addition to the broadening of STEM participation. As change takes place to improve the conditions in which Black and Brown youth engage with STEM for authentic and meaningful experiences that include computing and CT, program evaluation will be central to understanding any semblance of progress and assessing the realization of outcomes that truly matter for students.

Culturally Responsive Evaluation

The pervasiveness of culture in all aspects of human activity and endeavor is undeniable. Culture is defined as "a cumulative body of learned and shared behavior, values, customs and beliefs to a particular group or society" (Frierson et al., 2002, p. 63). The educational process of teaching and learning in formal and informal settings is a cultural activity (Stigler & Hiebert, 1998). Traditionally, STEM has engendered a false narrative of a culturally immune domain. In addition to being far from true, this false narrative serves a meritocratic exclusionary agenda detrimental to inclusive democratic principles that advance equity and social justice. The process of STEM education involves socialization to a culture of science, as well as culture-specific disciplinary norms and values. While the current ubiquitous nature of technology and its use can misleadingly signal the impression of a benign, race-neutral information age, the ways in which technology is organized, managed, and applied have profound cultural implications that adversely affect communities of color (Benjamin, 2019). In her groundbreaking book, *Race after Technology*, Benjamin (2019) describes "different forms of coded inequity" inherent in automated systems that influence every aspect of modern life, including healthcare, criminal justice, education, and consumer taste.

Culturally responsive evaluation positions culture in the center of evaluation and refutes the notion of culture-free evaluation (Hood et al., 2005; Hood et al., 2015; Hopson, 2009). Programs and interventions that are the object of evaluation function within culture-embedded environments. Evaluators who carry out program evaluation inherently approach systematic, disciplined inquiry through a cultural lens reflecting their epistemological, ontological, and theoretical orientations. Any program evaluation in broadening STEM participation claiming a blind eye to culture does a disservice to equity and undermines the integrity of meaningful, practical, and credible inquiry for marginalized communities of color. The American Evaluation Association (AEA), the national association of professional evaluators, includes culturally appropriate knowledge and skill sets among the essential competencies that professional evaluators should possess (AEA, 2011, 2018). Culturally responsive evaluation (CRE) considers culture of both the program evaluated and the subjects of the intervention to be important to the evaluative process (Frierson et al., 2002; Hood et al., 2015). A thorough description of CRE delineating its fundamental elements can be found in the 2015 *Handbook of Program Evaluation*.[4]

While we ascribe to CRE for evaluation of broadening STEM participation, the endeavor to amplify the voices of marginalized communities has particular poignancy for this chapter and in our professional experiences. An increasing body of research attests to the racialized experiences that students of color often encounter in STEM along the PreK through graduate school educational continuum (Joseph & Cobb, 2019; McGee, 2018, 2020; Nasir et al., 2009; Solorzano et al., 2000; Wilkins-Yel et al., 2019). Situated within the larger context of STEM education, the computer science education movement is not immune from structural inequalities and racialized disparities (Bobb & Brown, 2017; cf. Margolis et al., 2017). As we attend to the cultural environment of STEM teaching and learning by using theoretical orientations in alignment with broadening participation, appropriate instrumentation developed with diverse communities, and our own evaluative positioning as minority women of color, we foster inquiry that reflects the shared values and mutual trust with marginalized populations.

This inquiry in turn leads to greater reliability and validity of results. As an example, one of our projects encountered difficulty recruiting sufficient numbers of underrepresented minority students to participate in undergraduate research experiences. Subsequently, examination of recruitment strategies and missed participant targets suggested a number of programmatic changes were needed to reach and appropriately incentivize more students. Even with the need to refine recruitment strategies, our culturally responsive focus group protocols uncovered STEM identity and sense of belonging orientations that undergraduate minority students had internalized, which mitigated their conceptual alignment with the available opportunities to participate in undergraduate research. Students were aware of the opportunities, and they even knew about the availability of a sizeable stipend offered to students who participated in undergraduate research to offset the economic realities of the need to work outside of school. As they shared in our focus group sessions, students of color were not applying because their teaching and learning experiences in STEM reinforced the stereotype threat that they did not belong in a scientific research environment and, therefore, would not be successful. Students shared the reaction of literally shutting down when professors made announcements in class to encourage students to participate in funded research opportunities. One student described how she physically turned her chair around to face her back to the professor in class each time the announcement about available research opportunities was made. The Black female undergraduate just could not contend with the painful cumulative effect of negative messaging about a perceived inadequacy of her suitability for participating in scientific research. Once the project used culturally relevant counternarrative strategies (e.g., role models with whom students identified) to dismantle students' deficit-oriented schema, the participation of students of color in undergraduate research experiences increased and was sustained over the years.

Care must be exercised in the identification and selection of appropriate valid and reliable instruments in the conduct of any evaluation. The construct of multicultural validity is essential in culturally relevant evaluation to ensure congruence with culturally specific contexts (Hood et al., 2015). An example experienced by Toks, co-author of this text, illustrates the importance of multicultural validity. An evaluation in early childhood conducted several decades in the past examined the effects of a mathematics intervention on the performance of Black children in arithmetic. At the time, the principal investigator (PI) framed the evaluation through a deficit-oriented lens. Results from cognitive interviews during exploration of the addition of integers using manipulatives showed student mastery of the concept as the young students orally explained their thought process and problem-solving approach while completing the tasks. However, the formal oral assessment showed a lack of proficiency in the students' ability to add integers. Due to culturally responsive insights from Toks, additional investigation revealed the gap was caused not by a lack of student understanding but by misalignment of the oral assessment to the student learning context. Students were asked to name the *sum* of two integers when that arithmetic term had not been introduced or used with them to complete tasks with manipulatives.

Reflecting on the example of cultural relevance and the importance of arithmetic terms, we note the growing availability of instruments to use in evaluations focused on broadening STEM participation. Yet, it remains important to understand the context in which instruments have been piloted and field-tested as well as the experience of project evaluators who have used surveys and other protocols with diverse audiences. We continue to find the need to adapt instruments to increase alignment with CRE, and often seek additional information from the developer on relevant cultural design features (e.g., language, subscales). The fairly recent introduction of CT to the STEM education landscape means corresponding assessments and instruments may be limited. However, the computer science education community has increased its capacity to produce instruments responsive to diverse populations that measure CT knowledge, skills, and practice. There are several repositories and databases now available to identify appropriate instruments, tools, and protocols for use in broadening STEM participation evaluation with increasing attention to computing and CT.[5]

An essential practice we use to maintain grounding in cultural responsiveness is the application of relevant and appropriate theoretical and conceptual frameworks to design and carry out program evaluation. The theoretical frameworks that we have found useful for program evaluation and broadening STEM participation are associated with socio-cognitive theory and STEM education research. A fitting characterization of these would be in keeping with the term *broadening participation theory*, which often addresses social, cultural, and contextual aspects of STEM experienced by underrepresented minority students (Powell et al., 2018). Socio-cognitive theory and culturally relevant pedagogy through a lens of critical race theory have been particularly relevant. Brofenbrenner's (1979) ecological systems theory addresses the different environmental systems encountered by students as they navigate the world. His theory frames these environmental systems as environmental organizations that interact and contribute to an individual's encounters, development, beliefs, goals, and achievements. Environmental organizations are layered and include microsystems (friends, communities), mesosystems (immediate family dynamic), exosystems (gender, ethnicity, race, socioeconomic status), and chronosystems (situational conditions such as food insecurity, mobility, health of caretakers, etc.). Ecological systems theory frames our evaluative lens on students and their lived experiences in their immediate environments and the larger context of their lives.

We have relied on socio-cognitive theory (Bandura, 1977, 1986, 1997; Bandura & Schunk, 1981), attribution theory (Weiner, 1972, 2010), and locus of control theory (Rotter, 1990) to examine how interventions, including experiences to foster CT skills and competencies, contribute to participants' beliefs in their capacity for performance attainment. Limited self-efficacy characterized the beliefs of the Black student mentioned earlier in the example from one of our evaluations, who simply could not identify with seeing herself as capable of participating in undergraduate research. Self-efficacy is a predictor of future behavior that has contextual associations with motivation, locus of control, patterns of behavior, and the environment (Bandura, 1977, 1986). Finally, theories of culturally sustaining pedagogy (Paris, 2012), inclusive of culturally relevant pedagogy (Ladson-Billings, 1995, 2009a) and culturally responsive teaching (Gay, 2010), have made important contributions to STEM education with insightful relevance to broadening participation. In its rejection of deficit-oriented frameworks, culturally sustaining pedagogy insists upon empowering the agency of Black and Brown learners by building on students' capabilities, talents, and experiences. Culturally sustaining teaching and pedagogy involve "the ability to link principles of learning with deep understanding of (and appreciation for) culture" (Ladson-Billings, 2014, p. 77).

The broadening participation theories mentioned in this chapter represent those that we have found to be especially relevant to our work in STEM evaluation. The omission of any specific theory should not be interpreted with any significance for the purpose of this chapter. We encourage familiarity with a range of broadening participation theories, such as critical race theory (Davis & Jett, 2019; Ladson-Billings, 2009b; Ladson-Billings & Tate, 2006; O'Hara, 2020), intersectionality (Cho et al., 2013; Ireland et al., 2018), and identity development (Garciá et al., 2019; Kim et al., 2018) among others. The breadth and depth of scholarship in broadening participation now afford evaluators the opportunity to apply theory directly related to the lived experience of marginalized communities and students of color in the conduct of CRE.

Race, Ethnicity, and Data Collection

Common practice in the collection of self-reported race and ethnicity tends to rely on baseline data. Our use of surveys for data collection has uncovered complexities surrounding the stability of participant self-reporting on race and ethnicity, particularly among racialized minority groups. Limiting survey items on race and ethnicity to a single data collection point when working with diverse populations is problematic. By including items on race and ethnicity on each survey over the duration of the project, the opportunity for data loss on these important variables is minimized. The practice of including survey items on race and ethnicity at every point of data collection has been especially effective in our work in rural communities with multiethnic and multiracial populations. A similar pattern of self-reporting anomalies on race and ethnicity emerged in survey data across multiple points of administration with adult populations as well as minors. These participant self-reporting anomalies on race and ethnicity represented 15% of respondents in some cases—a percentage that presents a threat to reliability.

An emerging body of research suggests that individuals with multiracial and multiethnic backgrounds are presented with increased opportunity to change their self-reported racial and ethnic identity over time, a phenomenon usually not tracked in data collected at the institutional level (Masuoka, 2011; Tabb, 2015; Tabb et al., 2016). For instance, consider Clara in our opening vignette. Her identity was first characterized as Black and later listed as Afro-Latina. The change in her characterization was intentional to reflect Clara's identity development related to race and ethnicity. She most often selects Black when asked to identify her racial background and has been categorized as a Black female throughout her formal schooling experience. However, Clara's maternal great-grandfather migrated in the early 1900s to the United States from Chihuahua, Mexico. Over the years, Clara began to embrace her Mexican ancestry and would, therefore, select Hispanic on surveys, which usually appeared prior to race, and then select Black for her racial identity. She found herself still negotiating the meaning of her diverse ancestry and, therefore, at other times, she would determine to select Black exclusively as her racial and ethnic identity. She wondered, "Does Black in the US represent an amalgam of racial and ethnic groups anyway? So, when

I check Black, my Hispanic ancestry should be understood to be included. But by not checking Hispanic, I will be denying my Hispanic ancestry." She found herself in a quandary. Clara's approach to answering survey questions on race and ethnicity is illustrative of the complexities often encountered by individuals from multiracial and multiethnic backgrounds.

The proliferation of DNA testing in the search of discovering ancestry and ethnic origins contributes to new knowledge, sometimes unexpectedly, of an individual's racial and ethnic background. As a result, racial and ethnic identification are subject to change. Another contributing factor to inconsistencies in racial and ethnic self-reporting across multiple surveys may be conflicting identity dissonance derived from the pressure to assimilate or "fit-in" from external sources, such as schools, the political climate, and the schooling/community context. The complexities associated with self-identifying as an underrepresented minority student, especially for multiracial and multiethnic individuals, in predominantly majority environments may present conflicting challenges. Herman (2004) identified environmental factors related to the racial composition of the local community and school to be racial identification determinants for multiracial youth of Hispanic/Latinx ancestry. Although we use the term broadly, we recognize the rich ethnic and cultural diversity that exists in the Hispanic/Latinx U.S. population.

Survey items on race and ethnicity may be the first instance presented to youngsters in which they have the opportunity to formally reflect and report on identity. Our experience in collecting survey data on race and ethnicity in rural settings showed evidence of the process of identity development occurring for some students. Possibly, the survey catalyzed self-reflection on processing race and ethnicity in ways not previously experienced. Survey questions on race and ethnicity may have also prompted students to have conversations with parents and/or guardians and share information with their friends about their ancestry and racial/ethnic backgrounds. In one instance we recall when visiting schools in Wyoming following administration of a survey, a girl, who by appearance looked White, approached us when class was dismissed and said, "I am part American Indian." She then nonchalantly proceeded to join her classmates at lunch as part of her normal routine. The content of the class period that had just ended did not involve identity issues in any way. We had not spoken to the students about racial, ethnic, and/or cultural issues. We surmised the experience of completing the demographics questions on the survey prompted students to be mindful of their identity around race and ethnicity in ways they may not have experienced previously.

While racial and ethnic identity would seem to be more stable for adults, we have similarly encountered self-reporting inconsistencies across multiple teacher surveys, usually also in rural settings. Particularly in cases where individuals of multiracial and multiethnic backgrounds are in majority settings in more remote and isolated communities, we have found more variation in self-reporting on race and ethnicity across multiple surveys. According to conventional practice, it may seem redundant to include demographic questions on race and ethnicity on surveys that require multiple points of administration across the duration of a project. However, the absence of demographic questions on race and ethnicity in multiple surveys (e.g., pre and post) could potentially lead to data loss on variables critical to investigating broadening participation. It is possible for race and ethnic identity to fluctuate and evolve in these current times based on individuals' greater awareness and reflection, the conditions of the environment, and the prevailing context. We take all of these topics and concepts into consideration, as we reflect upon the second part of the chapter, which specifically addresses a broadening participation project that sought to engage underrepresented minority students in CT though robotics and game design.

Using Robotics and Game Design in Informal Settings to Foster Computational Thinking

We conducted the external evaluation for a four-year project funded by NSF's Innovative Technology Experiences for Students and Teachers (ITEST) program from October 1, 2013 through September 30, 2017. The project, uGame-iCompute, focused on providing students in elementary, middle, and junior high schools with opportunities to engage in CT through the use of robotics and game design with a special interest in reaching underrepresented minority students. The PI of the uGame-iCompute project, Jacqueline Leonard, a STEM education researcher, is the lead author of this book. The main goal of the uGame-iCompute project was to provide underrepresented and underserved students in Wyoming and Pennsylvania with opportunities to engage in CT through gaming and robotics. This chapter discusses the summative evaluation results of the uGame-iCompute project introduced in Chapter 1. The evaluation's summative outcomes in relationship to underrepresented minority students are presented. Formative results have been reported and disseminated elsewhere (Leonard et al., 2015, 2017, 2018; Mitchell et al., 2016).

The project design involved two primary components: (1) professional development of teachers through the offering of an eight-week online graduate course in CT and culturally responsive pedagogy, and (2) school-based student robotics and game design clubs facilitated by the teacher participants for a total of 40 to 50 contact hours with students each semester. Project participants learned to code using AgentSheets and AgentCubes applications in the context of game design. CT was embedded throughout the learning experience but also explicitly taught as content in the professional development course offering. The project team responsible for instruction was interdisciplinary and included a diverse group of mathematicians, mathematics educators, and computer scientists. The project theorized that underrepresented students' participation in culturally relevant experiences and CT offered in the context of computer-based Scalable Game Design and LEGO® robotics would have a positive influence on their STEM interest and attitudes. Implementation of culturally relevant pedagogy in the robotics and game design clubs took the form of students' use of place-based themes (e.g., mountains, frogs) and design features (e.g., music, artistic patterns), including representations from Indigenous communities that resonated with students' interests, backgrounds, and environments (Leonard et al., 2018).

The summative evaluation examined the extent to which the project produced positive participant change by posing the following questions to address project outcomes associated with the teachers and students:

1. To what extent did the project realize positive change in teachers' beliefs about CT, and culturally responsive teaching self-efficacy and outcome expectancy?
2. To what extent did the project realize positive change in students' technology self-efficacy and attitudes toward mathematics?

Setting and Participants

The out-of-school time (OST) clubs in robotics and game design took place in public schools in two different states. Participating schools in the Rocky Mountain region were located primarily in rural settings throughout the state of Wyoming. Schools in the mid-Atlantic region, considered a more urban setting, were based in school districts surrounding the Philadelphia area in Pennsylvania.

Teacher Participants

A total of 38 teachers participated in the project's professional development. These teachers organized and facilitated robotics and game design OST clubs in 29 elementary, middle, and junior high schools. The demographics of the participating teachers appear in Table 7.1. While most teachers were White and female, approximately 21.1% of the participating teachers were from underrepresented minority groups. The vast majority of the teacher participants were responsible for teaching a STEM-related subject during the formal school day.

Student Participants

In total, 228 students in elementary, middle, and junior high school participated in the evaluation of the project, although more students actually attended the OST robotics and game design clubs over the four-year project duration. Only participants with written parental consent and student assent were included in the evaluation. The demographics of students participating in the evaluation appear in Table 7.2. Almost two-thirds of the student participants were males. The total student sample was rather diverse ethnically. According to project-level

TABLE 7.1 Number of Participating Teachers by Race, Gender, and Subject Area

Teacher Demographics	#Analytic Sample
Race★	
Asian	1
Black	2
Latinx	6
Native American	2
White	29
Two or More Races	1
Gender	
Female	30
Male	8
Program★	
Elementary	4
Math	10
Science	4
Special Education	2
STEM	5
Computer Science	1
Technology	7
Other	3

★ Some individuals selected more than one category.

TABLE 7.2 Number of Participating Students by Race and Gender

Student Demographics	#Analytic Sample
Race/Ethnicity★	
Asian	6
Black	38
Latinx	24
Native American	30
White	137
Two or More Races	50
Missing Information	2
Gender	
Female	83
Male	144
Missing Information	1

★ Some individuals selected more than one category.

data, approximately 30% of participating students were from underrepresented minority groups, which among them included 19% Black, 9% Hispanic/Latinx, and 1% Native American (Leonard, 2017). Most of the schools participating in the project were designated as Title 1, indicating student enrollment with large concentrations of youngsters who were eligible for free or reduced lunch.

Theoretical Framing

The theoretical framework that grounds the summative evaluation is Bandura's social cognitive theory of self-efficacy (Bandura, 1977, 1986). Self-efficacy theory is based on the beliefs of individuals that they possess the capacity to achieve certain behaviors, goals, and perform at certain levels. As a result, self-efficacy argues that individuals will participate in activities in which they believe their competence aligns with performance expectations. The sources of self-efficacy theory correspond to teaching and learning in the following way. Enactive mastery involves prior performance in accomplishing tasks. Therefore, the extent to which students experience success in mastering content will impact their self-efficacy. Vicarious experience as a source of self-efficacy involves opportunities to observe others who successfully complete tasks. In the context of broadening participation, this could involve peers or role models of similar racial or ethnic background experiencing STEM success. A third source of self-efficacy is the component of verbal persuasion. This source finds expression in social interactions within the dynamics of teaching and learning corresponding to encouragement, feedback, and positive discourse around STEM. The final source is physiological implications wherein the affective domain is energized for performance tasks that support success with an inverse relationship that is associated with performance expectations that may cause stress due to stigmatizing expectations around failure.

Self-efficacy is a predictor of performance and behavior. Positive change in self-efficacy represents confidence and a motivation to succeed, which holds promise for improved performance.

Methods

The evaluation relied on quantitative methods where surveys were the primary data source. Additional data sources included observations of the OST gaming and robotics clubs using the Dimensions of Success observation tool (Noam et al., 2014; Papazian et al., 2013). The results associated with observations of the OST gaming and robotics clubs are not included in this chapter and are reported elsewhere (Leonard et al. 2017; Mitchell et al., 2016). Administration of surveys occurred at the beginning and end of each semester associated with project participation. Online surveys were administered to the teachers at two data collection points—prior to the beginning and at the end of their professional development experience. The two data collection points for administration of the teacher surveys allowed for pre- and post-values to measure change. Teachers administered surveys on self-efficacy and attitudes to participating students prior to the start of the OST robotics and game design clubs and again at the end of the intervention as a pre-post measure.

Importantly, the launch of the project occurred when survey instruments to measure CT were sparse or in the development stage. At the time, most assessment revolved around the context of engagement with CT activities (Gardner et al., 2012; Koh et al., 2014). Curriculum development to foster CT and accompanying assessments were also in the early stages. However, we are encouraged by greater availability of instruments and tools that allow for more direct measurement of CT. The current ITEST project entitled the Bessie Coleman Project, which builds on the uGame-iCompute project, completed survey adaptation to derive CT self-efficacy indicators, which should be useful to the field (Jordan, 2020).

Teacher Data Sources

The Computing Attitudes Questionnaire (Yadav et al., 2014) was used to measure teacher attitudes in CT. Yadav et al. (2014) adapted an existing instrument, the computer science attitudes survey for science and engineering students to develop the Computing Attitudes Questionnaire (Hoegh & Moskal, 2009). Each construct in the original instrument has a Cronbach alpha greater than 0.70, indicating good reliability. The 21-item, four-point Likert scale (1 = strongly disagree, 2 = disagree, 3 = agree, 4 = strongly agree) on the Computing Attitudes Questionnaire served as the basis for rating five specific constructs: (a) understanding CT (e.g., *Computational thinking involves thinking logically to solve problems*), (b) comfort with computing (e.g., *I can learn to understand computing concepts*), (c) interest in computing (e.g., *I think computer science is interesting*), (d) incorporating CT into the classroom (e.g., *Computational thinking can be incorporated in the classroom by allowing students to problem solve*), and (e) the role of computing in careers (e.g., *I expect that learning computing skills will help me to achieve my career goals*).

To investigate culturally responsive teaching, we used scales developed by Siwatu (2007), originally developed for use with preservice teachers. The culturally responsive teaching self-efficacy (CRTSE) and culturally responsive teaching outcome expectancy (CRTOE) scales allowed us to evaluate teacher self-efficacy in culturally responsive instructional delivery settings. According to the developer, the scales are useful for determining the effectiveness of programs that seek to foster culturally responsive teaching skills and competencies. The CRTSE and CRTOE scales were based on a 0–100 response format for greater response discrimination because the developer believed the traditional range of Likert formatted scales were psychometrically limited (Siwatu, 2007). The 40-item CRTSE scale measures teacher self-efficacy in executing teaching practices and tasks aligned with culturally responsive pedagogy (e.g., *I am able to build a sense of trust in my students; I am able to use my students' cultural background to make learning meaningful*). The 26-item CRTOE scale focuses on teachers' beliefs that culturally responsive teaching practices will lead to positive teaching and learning outcomes (e.g., *Using culturally familiar examples will make learning new concepts easier; students' self-esteem can be enhanced when their cultural background is valued by the teacher*).

The items from each instrument were consolidated into a single teacher survey. The pre-post survey data were analyzed using *t*-tests to determine the extent to which the intervention contributed to significant differences in teachers' attitudes toward CT, CRTSE, and CRTOE. Teacher subgroup analysis by gender was also examined. A subgroup analysis by teachers' race and ethnicity was not possible, given the small number of teachers with pre-post survey data from underrepresented minority groups.

Student Data Sources

The evaluation drew from two specific instruments to collect student survey data. The Self-Efficacy in Technology and Science (SETS) instrument, developed by Ketelhut (2010), was selected to examine students' self-efficacy in technology. The instrument was designed for use with middle school students within the context of technology-based environments such as game design. The evaluation focused on three of the six self-efficacy subscales from SETS, which included videogaming self-efficacy (belief in the ability to perform effectively in videogaming), computer gaming self-efficacy (belief in the ability to perform effectively in computer gaming), and problem solving using the computer self-efficacy (belief in the ability to problem solve when using the computer). The videogaming self-efficacy subscale consisted of 8 items ($\alpha = 0.93$), (e.g., *I am sure I can learn any videogame*); the computer gaming self-efficacy subscale consisted of 5 items ($\alpha = 0.84$), (e.g., *I am very good at building things in simulation*); and the problem solving using the computer subscale consisted of 5 items ($\alpha = 0.79$), (e.g., *I can find information on the web by using a search engine*). Cronbach's alpha for each subscale reflected sufficient reliability.

A second survey, Student Attitudes Toward Science, Technology, Engineering and Math (S-STEM) developed by Unfried et al. (2015), was used to measure student attitudes toward mathematics and engineering/technology. The attitudes toward mathematics subscale consisted of 8 items ($\alpha = 0.90$), (e.g., *I can get good grades in math*), and the attitudes toward engineering/technology subscale consisted of 9 items ($\alpha = 0.90$), (e.g., *I am good at building and fixing things*). Both subscales were within the range of sufficient reliability.

The subscale items from all instruments were consolidated into one survey. The pre-post survey data were analyzed using *t*-tests to determine the extent to which the intervention contributed to significant differences in underrepresented minority students' self-efficacy and attitudes. In cases of statistical significance, Cohen's *d* was used to determine effect size (Cohen, 1988). Student subgroups by gender and race/ethnicity were also examined. While students from Indigenous ancestry participated in the evaluation, their numbers were insufficient to include as a separate subgroup in the pre-post survey analysis. Due to the small sample size for some subgroups ($n < 20$), Hedges correction was applied to mitigate bias (Lakens, 2013). In every case, these effect size estimates were equivalent. Therefore, effect size reporting is based on Cohen's d.

The student survey results reported in this chapter are limited to underrepresented minority participants. The term *underrepresented minority* (URM) refers to racial and ethnic groups disproportionately underrepresented in STEM fields compared to their representation in the U.S. population. Participants self-reporting their racial and/or ethnic background in any of the following categories were considered URM students: Black/African American, Hispanic/Latinx, Native American/Alaska Native, or "two or more races" wherein at least one corresponded to the preceding racial and ethnic categories. Due to the racial and ethnic demographics of participant locations in Wyoming and the greater Philadelphia area, the survey did not include Native Hawaiian/Other Pacific Islander as a racial and/or ethnic response choice.

Results

Summative in nature, the evaluation results of the uGame-iCompute project reflect the project period corresponding with full implementation over four semesters beginning in fall 2014 and ending at the close of spring 2016. Project results accomplished on an annual basis and/or by term(s) are reported elsewhere (Leonard et al., 2015, 2016, 2018; Newton et al., 2020). Teacher participants completed semester-long professional development equivalent to approximately 30 hours to learn computer programming using AgentSheets and AgentCubes, and culturally responsive pedagogy to prepare them to organize and facilitate robotics and game design during OST clubs. The content of the professional development sessions included teacher instruction in the activities that students would experience in the robotics and game design clubs (see Chapter 6). Students voluntarily attended the robotics and game design clubs organized by the project team and facilitated by the teacher participants. Club participation involved 30 to 40 hours of content in an OST environment, typically afterschool. Student attendance records were collected by the teacher participants and submitted to the project team. However, inconsistencies in attendance reporting prevented the rates of student participation or exposure to treatment from being included in the analysis.

Teacher Change

Teacher participants completed pre-post surveys on culturally responsive teaching and attitudes toward CT. Commencing with the fall 2014 semester, surveys were administered to four cohorts at the start and end of each term that corresponded with teachers' formal professional development provided by the project. The teacher survey results were analyzed using a paired t-test. The participating teachers made significant positive gains in their CRTSE. As shown in Table 7.3, teachers' mean CRTSE scores increased significantly from pretest (80.50 [$SD =$ 11.06]) to posttest (85.10 [$SD = 13.51$]) with a relatively small effect size ($d =$ 0.373).

Survey results related to the CT subscales show significant teacher gains in teacher attitudes toward integrating CT into the classroom. As shown in Table 7.4, the teachers' mean CT classroom integration scores increased significantly from pretest (3.20 [$SD = 0.31$]) to posttest (3.42 [$SD = 0.29$]) with a medium effect size ($d = 0.7238$).

Additional subgroup analyses uncovered gender differences in teacher attitudes toward CT. Female teachers made significant gains in their attitudes toward integrating CT into classroom practice. As shown in Table 7.5, the mean CT

TABLE 7.3 *T*-test Teacher Culturally Responsive Teaching Self-Efficacy and Outcome Expectancy Survey Results

CONSTRUCT	N	Pretest (SD)	Posttest (SD)	T-Value	Probability
Culturally Responsive Teaching Self-Efficacy	38	80.50 (11.06)	85.10 (13.51)	−2.753	0.014★
Culturally Responsive Teaching Outcome Expectancy	38	86.02 (13.0)	88.56 (9.61)	−1.885	0.067

★ $p \leq 0.05$

TABLE 7.4 *T*-test Teacher Survey Results: CT Constructs ($n = 37$)

CONSTRUCT	Pretest (SD)	Posttest (SD)	T-Value	Probability
Understanding CT	3.11 (0.11)	3.22 (0.07)	−1.671	0.103
Comfort with Computing	3.45 (0.21)	3.42 (0.22)	0.563	0.577
Interest in Computing	3.16 (0.39)	3.17 (0.53)	−0.075	0.941
CT Classroom Integration	3.20 (0.31)	3.42 (0.29)	−2.525	0.016★
CT Career Relevance	3.33 (0.23)	3.40 (0.28)	−0.846	0.403

★ $p \leq 0.05$

TABLE 7.5 *T*-test Female Teachers' Survey Results: Computational Thinking Constructs ($n = 29$)

CONSTRUCT	Pretest (SD)	Posttest (SD)	T-Value	Probability
Understanding CT	3.10 (0.12)	3.19 (0.08)	−1.08	0.289
Comfort with Computing	3.50 (0.21)	3.45 (0.12)	0.845	0.405
Interest in Computing	3.21 (0.47)	3.23 (0.58)	−0.281	0.781
CT Classroom Integration	3.24 (0.33)	3.48 (0.28)	−2.317	0.028*
CT Career Relevance	3.37 (0.23)	3.43 (0.34)	−0.688	0.497

★ $p \leq 0.05$

TABLE 7.6 *T*-test Male Teachers' Survey Results: Computational Thinking Constructs (*n* = 8)

CONSTRUCT	Pretest (SD)	Posttest (SD)	T-Value	Probability
Understanding CT	3.13 (0.09)	3.31 (0.03)	−3.000	0.021★
Comfort with Computing	3.27 (0.19)	3.31 (0.27)	−0.314	0.763
Interest in Computing	3.00 (0.07)	2.94 (0.37)	0.239	0.818
CT Classroom Integration	3.06 (0.25)	3.19 (0.28)	−1.000	0.351
CT Career Relevance	3.20 (0.23)	3.30 (0.08)	−0.468	0.654

★ $p \leq 0.05$

classroom integration scores of female teachers increased significantly from pretest (3.24 [*SD* = 0.33]) to posttest (3.48 [*SD* = 0.28]) with a relatively large effect size (*d* = 0.7894). We did make note of a slight decline in female teachers' comfort with computing, albeit not statistically significant, which may be related to the presence of some dissonance associated with learning new subject matter.

While male teachers did make slight gains in their attitudes toward CT classroom integration, the change was not significant. Though the gains in male teachers' understanding of CT were significant. As shown in Table 7.6, the understanding CT scores of male teachers increased significantly from pretest (3.13 [*SD* = 0.09]) to posttest (3.31 [*SD* = 0.03]) with a relatively large effect size (*d* = 0.7638). We also note the relative stability in male teachers' comfort with computing compared to their female counterparts who experienced a slight decline. The observation of significant gains in male teachers' understanding CT, which exceeded the post-scores of their female counterparts, would suggest their comfort with computing would resemble a similar pattern. However, the pre-and post-comfort levels with computing scores of male teachers fell below their female counterparts. It would be important to explore potential gender differences and teachers' CT beliefs through additional studies and evaluation.

Student Change

Students completed surveys on self-efficacy and attitudes prior to and at the end of their participation in semester-long robotics and game design clubs organized each term starting with the fall 2014 semester through spring 2016. Student participants comprised a unique cohort each term. Pre-post data for individual students were limited to one term of an OST robotics and gaming experience. The student survey results reflect outcomes for underrepresented minority participants and were analyzed using a paired *t*-test.

Although URM students' videogaming self-efficacy was positive over the course of the intervention, these gains among the broad subgroup categories were not statistically significant. (See Table 7.7.) Generally, the videogaming

TABLE 7.7 *T-test Videogaming Self-Efficacy by Student Subgroup*

Student Subgroup	N	Pretest (SD)	Posttest(SD)	T-Value	Probability
Underrepresented Minorities	82	4.06 (0.48)	4.04 (0.69)	0.245	0.807
Black Students	38	4.16 (0.41)	4.21 (0.56)	−0.499	0.620
Latinx Students	24	3.88 (0.28)	4.08 (0.39)	−1.840	0.079

TABLE 7.8 *T-test Student Videogaming Self-Efficacy by Gender[†]*

CONSTRUCT	N	Pretest (SD)	Posttest (SD)	T-Value	Probability
Underrepresented Female Minorities	38	3.94 (0.37)	3.85 (0.65)	0.967	0.337
Underrepresented Male Minorities	43	4.17 (0.56)	4.21 (0.68)	−0.311	0.757
Black Female Students	20	3.94 (0.37)	3.85 (0.65)	0.967	0.337
Black Male Students	18	4.33 (0.43)	4.43 (0.26)	−0.758	0.459
Latina Students	11	3.79 (0.33)	3.70 (0.26)	0.921	0.379
Latino Students	13	3.95 (0.26)	4.40 (0.29)	−2.896	0.013★

★ $p \leq 0.05$
† Some students did not disclose their gender and race/ethnicity.

TABLE 7.9 *T-test Computer Gaming Self-Efficacy by Student Subgroup*

Student Subgroup	N	Pretest (SD)	Posttest (SD)	T-Value	Probability
Underrepresented Minorities	82	3.92 (0.40)	4.04 (0.44)	−1.989	0.050★
Black Students	38	4.12 (0.41)	4.16 (0.37)	−0.446	0.658
Latinx Students	24	3.55 (0.23)	3.86 (0.45)	−3.000	0.006★★

★ $p \leq 0.05$
★★ $p \leq 0.01$

self-efficacy pre- and post-scores of Latinx students were lower than that of their Black counterparts.

When examining URM results by gender, statistically significant videogaming self-efficacy was evident for Latino male students. As shown in Table 7.8, the videogaming self-efficacy scores of Latino males increased significantly from pretest (3.95 [$SD = 0.26$]) to posttest (4.40 [$SD = 0.29$]) with a large effect size ($d = 0.8628$). As a pattern, male students' pre- and post-videogaming self-efficacy scores exceeded that of their female counterparts.

All **underrepresented minority** students made significant gains in computer gaming self-efficacy. As shown in Table 7.9, URM students' computer gaming self-efficacy scores increased significantly from pretest (3.92 [$SD = 0.40$]) to posttest (4.04 [$SD = 0.44$]) with minimal effect size ($d = 0.1721$). Among the URM subgroups, Latinx students' computer gaming self-efficacy scores increased significantly from pretest (3.55 [$SD = 0.23$]) to posttest (3.86 [$SD = 0.45$]) with medium effect size ($d = 0.5364$).

TABLE 7.10 *T*-test Student Computer Gaming Self-Efficacy by Gender†

Gender Subgroup	N	Pretest (SD)	Posttest (SD)	T-Value	Probability
Underrepresented Female Minorities	38	3.84 (0.30)	3.95 (0.48)	−1.346	0.186
Underrepresented Male Minorities	42	4.00 (0.48)	4.11 (0.41)	−1.451	0.154
Black Female Students	20	4.00 (0.31)	4.08 (0.46)	−0.856	0.403
Black Male Students	18	4.27 (0.51)	4.25 (0.29)	0.151	0.882
Latina Students	11	3.51 (0.27)	3.69 (0.43)	−1.656	0.271
Latino Students	13	3.58 (0.22)	4.01 (0.46)	−3.062	0.010★

★ $p \leq 0.05$
† Some students did not disclose their gender and race/ethnicity.

TABLE 7.11 *T*-test Problem Solving Using the Computer Self-Efficacy by Student Subgroup

Student Subgroup	N	Pretest (SD)	Posttest (SD)	T-Value	Probability
Underrepresented Minorities	82	3.97 (0.50)	4.05 (0.60)	−0.986	0.327
Black Students	38	4.16 (0.55)	4.15 (0.65)	0.134	0.894
Latinx Students	24	3.77 (0.35)	4.12 (0.47)	−2.297	0.031★

★ $p \leq 0.05$

Significant computer gaming self-efficacy gains by gender were limited to Latino males. As shown in Table 7.10, the computer gaming self-efficacy scores of Latino males increased significantly from pretest (3.58 [$SD = 0.22$]) to posttest (4.01 [$SD = 0.46$]) with a medium effect size ($d = 0.7323$).

The problem solving using the computer self-efficacy subscale scores revealed that Latinx students were the only URM student subgroup with statistically significant gains. As shown in Table 7.11, the problem solving using the computer self-efficacy scores of Latinx students increased significantly from pretest (3.77 [$SD = 0.35$]) to posttest (4.12 [$SD = 0.47$]) with a medium effect size ($d = 0.5468$). While the pretest problem solving using the computer self-efficacy score of Latinx students lagged behind their Black counterparts, their posttest scores were similar.

Gender differences in students' problem solving self-efficacy did not emerge. As shown in Table 7.12, the problem solving self-efficacy scores by gender showed no statistically significant gains.

The student attitudes toward mathematics subscale scores also show that Latinx students were the only URM subgroup with significant gains. As shown in Table 7.13, the mathematics attitudes scores of Latinx students increased significantly from pretest (3.83 [$SD = 0.55$]) to posttest (4.06 [$SD = 0.63$]) with a small effect size ($d = 0.2950$).

When examining gender differences in students' attitudes toward mathematics, Latino males appeared to be the only subgroup with statistically significant

TABLE 7.12 *T*-test Student Problem Solving Using the Computer Self-Efficacy by Gender[†]

Gender Subgroup	N	Pretest (SD)	Posttest (SD)	T-Value	Probability
Underrepresented Female Minorities	36	4.05 (0.60)	4.09 (0.82)	−0.398	0.693
Underrepresented Male Minorities	42	3.92 (0.43)	4.01 (0.45)	−0.656	0.516
Black Female Students	20	4.14 (0.66)	4.18 (1.00)	−0.311	0.759
Black Male Students	18	4.19 (0.47)	4.11 (0.29)	0.417	0.682
Latina Students	11	3.85 (0.44)	4.11 (0.69)	−1.218	0.251
Latino Students	13	3.69 (0.28)	4.12 (0.33	−1.923	0.079

[†] Some students did not disclose their gender and race/ethnicity.

TABLE 7.13 *T*-test Attitudes Toward Mathematics by Student Subgroup

CONSTRUCT	N	Pretest (SD)	Posttest (SD)	T-Value	Probability
Underrepresented Minorities	82	3.96 (0.61)	3.99 (0.67)	−0.493	0.632
Black Students	38	3.98 (0.76)	4.01 (0.73	−0.277	0.783
Latinx Students	24	3.83 (0.55)	4.06 (0.63)	−2.819	0.010★

★ $p \leq 0.05$

TABLE 7.14 *T*-test Student Attitudes Toward Mathematics by Gender[†]

CONSTRUCT	N	Pretest (SD)	Posttest (SD)	T-Value	Probability
Underrepresented Female Minorities	38	3.95 (0.51)	3.97 (0.63)	−0.185	0.854
Underrepresented Male Minorities	43	3.96 (0.70)	4.01 (0.73)	−0.492	0.624
Black Female Students	20	3.97 (0.70)	4.06 (0.58)	−0.782	0.444
Black Male Students	18	4.00 (0.87)	3.95 (0.94)	0.311	0.759
Latina Students	11	3.57 (0.34)	3.80 (0.74)	−1.663	0.127
Latino Students	13	4.05 (0.66)	4.28 (0.47)	−2.276	0.042★

★ $p \leq 0.05$
[†] Some students did not disclose their gender and race/ethnicity.

gains. As shown in Table 7.14, the mathematics attitudes of Latino males increased significantly from pretest (4.05 [SD = 0.66]) to posttest (4.28 [SD = 0.47]) with a small effect size (d = 0.4002). Moreover, the pre- and post-scores of Latino males on the mathematics attitudes construct exceeded their gender subgroup counterparts.

URM students' attitudes toward engineering/technology declined significantly, albeit their scores remained in the positive range. As shown in Table 7.15, the engineering/technology attitudes of URM students decreased significantly from pretest (4.20 [SD = 0.49]) to posttest (4.05 [SD = 0.66]) with a small effect size (d = 0.1987).

Among the gender subgroups, male URM students experienced statistically significant declines in their attitudes toward engineering/technology. As shown

TABLE 7.15 *T*-test Engineering & Technology Attitudes by Student Subgroup

CONSTRUCT	N	Pretest (SD)	Posttest (SD)	T-Value	Probability
Underrepresented Minorities	82	4.20 (0.49)	4.05 (0.66)	2.243	0.028*
Black Students	38	4.34 (0.48)	4.17 (0.52)	1.948	0.059
Latinx Students	24	4.05 (0.36)	4.04 (0.70)	0.109	0.915

* $p \leq 0.05$

TABLE 7.16 *T*-test Student Attitudes Toward Engineering and Technology by Gender[†]

CONSTRUCT	N	Pretest (SD)	Posttest (SD)	T-Value	Probability
Underrepresented Female Minorities	38	3.99 (0.55)	3.91 (0.66)	0.946	0.350
Underrepresented Male Minorities	43	4.38 (0.37)	4.17 (0.63)	2.081	0.044*
Black Female Students	20	4.09 (0.58)	4.06 (0.53)	0.349	0.731
Black Male Students	18	4.61 (0.23)	4.29 (0.50)	2.140	0.047*
Latina Students	11	3.86 (0.29)	3.70 (0.70)	0.675	0.515
Latino Students	13	4.21 (0.38)	4.32 (0.32)	−0.618	0.548

* $p \leq 0.05$
[†] Some students did not disclose their gender and race/ethnicity.

in Table 7.16, URM male students' engineering/technology attitude scores decreased significantly from pretest (4.38 [$SD = 0.37$]) to posttest (4.17 [$SD = 0.63$]) with a relatively small effect size ($d = 0.4130$). Additionally, the engineering/technology attitude scores of Black male students decreased significantly from pretest (4.61 [$SD = 0.23$]) to posttest (4.29 [$SD = 0.50$]) with a large effect size ($d = 0.8311$). Despite these significant declines, the posttest engineering/technology scores remained in the positive range. Latino males were the only subgroup to show an increase in attitudes toward engineering/technology. However, their scores were not statistically significant.

Limitations of the Evaluation

It is important to recognize the small sample size in the subgroup analysis, which leads us to interpret the results with a degree of caution. The voluntary nature of participation may have contributed to selection bias. Teachers and students at some of the Wyoming sites were predisposed to STEM experiences in robotics, while game design was unique. As a result, participants may have been more susceptible to responding positively to the experience. This may not have been as much of a factor with the teacher participants who were incentivized with the provision of course credit. Finally, format and data collection inconsistencies associated with the attendance records submitted by the participating teachers resulted in unusable data. Therefore, it was not possible to track student participation in the OST robotics and game design clubs, which prevented an analysis of participant engagement by treatment dosage.

Discussion

Progress in broadening STEM participation is often characterized by a myopic focus on increasing the quantity of underrepresented students or "moving the needle" as some would call it. Obviously, attracting and retaining more underrepresented minority students in STEM is necessary, but it is not sufficient to realize the true 21st-century potential that broadening STEM participation demands. The ways in which underrepresented minority students experience STEM fundamentally shape how they internalize their place, role, and future in relation to STEM. Our discussions of culturally relevant evaluation and broadening participation theories are offered to examine the lived STEM experiences of underrepresented minority students in the conduct of program evaluation.

The summative results of the NSF-funded uGame-iCompute ITEST project reported in this chapter provide evidence of the value of a robotics and game design intervention to underrepresented minority students. The significant self-efficacy and attitudinal gains of the underrepresented minority student participants are positive indicators of quality experiences in robotics and game design that had meaning in their perceptions related to motivation and confidence around STEM. The in-service component produced significant gains in teachers' culturally responsive teaching self-efficacy. The self-efficacy and attitudinal gains experienced by students and teachers reflect internal change derived from lived experience. The predictive relationship of self-efficacy gains to positive participant behavior supports the inference that teachers translated their significant gains in CRTSE in the facilitation of the OST gaming and robotics clubs. Evidence of culturally responsive teaching practices in the OST robotics and game design clubs that took the form of programming design features has been reported elsewhere (Leonard et al., 2015, 2018).

The interplay of culture and STEM through the conduct of robotics and game design provided the dynamics for teachers and students to make significant gains as a result of participants' engagement in the project. Notably, the project demonstrated the malleability of teacher attitudes in CT, including its integration in the classroom. The gains in teachers' CT attitudes by gender, with only male teachers experiencing significant gains in CT understanding, warrant additional research and program evaluation. The significant gains in teachers' attitudes toward integrating CT in classroom practice are promising. They suggest the transfer of in-service learning to practice from informal to formal settings is possible. This transfer relationship of CT practices between informal and formal settings should be investigated further.

Significant gains were made in underrepresented minority students' computer gaming self-efficacy, which aligned with the strong computer programming emphasis of the OST robotics and game design clubs. Among the racial and ethnic subgroups, Latinx students made significant gains on most of the subscales, including computer gaming self-efficacy, problem solving using the computer self-efficacy, and attitudes toward mathematics. When data were disaggregated by race and gender, Latino males were the sole subgroup that experienced significant gains for any construct. This finding suggests that while Latino male students did not necessarily outscore their peers, they most likely benefited the most from the experience and exposure. Underrepresented minority females, Black and Latina female students, did not make significant gains in any area. Thus, intersectionality (i.e., attention to race, gender, class, and other variables) is an important aspect of CRE (Rankin et al., 2020) and the ways in which STEM is experienced.

While underrepresented minority students' attitudes toward engineering/technology decreased significantly, with the exception of Latino boys, their post-test scores remained positive at the end of the intervention. A similar pattern was seen among underrepresented minority males and Black males. Although the factors contributing to these subgroup differences are unclear, contributions from intersectionality may shed additional light. Additional studies and program evaluation to examine associated factors of minority racial and ethnic subgroup differences in the context of robotics, game design, and CT are needed. Finally, Latinx students, particularly Latino males, saw significant gains in their attitudes toward mathematics. Although significant gains in mathematics attitudes were not seen by any other subgroup, most pre-post changes were positive. The transfer of CT through robotics and game design in informal settings to attitudes toward mathematics may be an area in need of further study.

Summary

The significant results from the robotics and gaming clubs for the Latinx students, particularly boys, are noteworthy. The majority of the Latinx participants attended predominantly White schools designated as Title I and located in remote rural areas in Wyoming. High poverty schools in rural settings often face challenges in overcoming the digital divide to provide technological access and connectivity (Beesley, 2011). The opportunity to experience gaming and robotics may have been a novel experience for the Latinx students. A nascent body of research suggests that computing and CT practices that feature visual-centric forms of programming, such as Scratch and AgentSheets, may facilitate multiple points of entry for Latinx and multilingual students (Jacob et al., 2018). Additionally, students experienced gaming and robotics through culturally responsive facilitation (Leonard et al., 2017: Mitchell et al., 2016). In their overview of research on broadening participation of Latinx students in computer science, Denner et al. (2017) highlight the responsive component of cultural relevance as an important factor. It is likely that an element of novelty coupled with multiple points of entry for engagement with computational practices, supported by culturally responsive pedagogy through gaming and robotics, contributed to the significant gains made by Latinx male students. Efforts at making inroads in broadening participation will need to be informed by additional scholarship on the participation of language learners in computing and computer science education (Jacob et al., 2018).

The summative outcomes from the external evaluation of the uGame-iCompute project bode well for broadening STEM participation by engaging diverse students in CT through gaming and robotics. We encourage continued opportunities for underrepresented minority students to participate in CT contexts similar to gaming and robotics. Our findings identified several areas in need of further research. As similar interventions and projects are implemented, it will be important to examine subgroup differences that emerged in our findings related to race, gender, and intersectionality, in more depth. The lived STEM experiences of countless underrepresented minority students who possess limitless potential to be the next generation of scientific innovators deserve no less.

Notes

1 The science and engineering (S&E) workforce includes occupations related to the life sciences, computer science, mathematics, physical science, social science, and engineering (National Science Board, 2018).
2 NSF has two merit review criteria: intellectual merit and broader impacts. To learn more about broadening participation as part of the NSF merit review process, see Broadening Participation at the National Science Foundation: A framework for action, August 2008 (https://www.nsf.gov/od/broadeningparticipation/nsf_framework-foraction_0808.pdf).
3 More information on the important role of CEOSE in addressing broadening participation at the NSF can be found in Advancing evaluation of STEM efforts through attention to diversity and culture in *New Directions for Evaluation* by Mertens & Hopson (2006).
4 For additional information on culturally responsive evaluation, the following sources are suggested: Culturally responsive evaluation as a form of critical qualitative inquiry by Michelle Bryan and Ashlee Lewis (2019); Continuing the Journey to Reposition Culture and Cultural Context in Evaluation Theory and Practice by S. Hood, R. Hopson, and H. Frierson (2015); and Culturally Responsive Evaluation Theory, Practice, and Future Implications by S. Hood, R. K. Hopson and K. E. Kirkhart (2015), in K. E. Newcomer, H. P. Hatry and J. S. Wholey (Eds.), *Handbook of practical program evaluation* (4th ed.), pp. 334–391.
5 Several repositories have been funded by NSF including STELAR (stelar.educ.org/resources/instruments) and CAISE (informalscience.org/evaluation/evaluation-tools-instruments) as examples. A recently developed evaluation repository is available through a coalition of partners that includes the American Evaluation Association's STEM Topical Interest Group, the Google-oriented CS Impact Network, and Oak Ridge Associated Universities accessible at https://www.orau.org/research-reviews-evaluations/stem-evaluation/evaluation-repository.html.

References

American Evaluation Association [AEA]. (2011). *Statement on cultural competence in evaluation*. AEA. https://www.eval.org/ccstatement

American Evaluation Association [AEA]. (2018). *AEA Evaluator Competencies*. AEA. https://www.eval.org/page/competencies

Bandura, A. (1977). Self-efficacy: Toward a unifying theory of behavioral change. *Psychological Review, 84*(2), 191–215.

Bandura, A. (1986). *Social foundations of thought and action: A social cognitive theory.* Prentice-Hall.

Bandura, A. (1997). *Self-efficacy: The exercise of control.* W. H. Freeman.

Bandura, A., & Schunk, D. H. (1981). Cultivating competence, self-efficacy, and intrinsic interest through proximal self-motivation. *Journal of Personality and Social Psychology, 41*(3), 586–598.

Beesley, A. (2011). Keeping rural schools up to full speed. *T.H.E. Journal, 38*(9), 26–27.

Bellman, S., Burgstahler, S., & Chudler, E. (2018). Broadening participation by including more individuals with disabilities in STEM: Promising practices from an engineering research center. *American Behavioral Scientist, 62*(5), 645–656.

Benjamin, R. (2019). *Race after technology: Abolitionist tools for the new Jim Code.* Policy Press.

Bobb, K., & Brown, Q. (2017). Access, power, and the framework of a CS education ecosystem. In Y. Rankins & J. Thomas (Eds.), *Moving students of color from consumers to producers of technology* (pp. 245–260). IGI Global.

Brofenbrenner, U. (1979). *The ecology of human development: Experiments by nature and design.* Harvard University Press.

Cho, S., Crenshaw, K. W., & McCall, L. (2013). Toward a field of intersectionality studies: Theory, applications, and praxis. *Signs, 38*(4), 785–810.

Cohen, J. (1988). *Statistical power analysis for the behavioral sciences.* Routledge. https://doi.org/10.4324/9780203771587

Committee on Equal Opportunities in Science and Engineering [CEOSE]. (2014). *Broadening participation in America's science and engineering work force: The 2013-2014 Biennial reports to congress.* National Science Board, National Science Foundation. https://www.nsf.gov/od/oia/activities/ceose/reports/ceose2004report.pdf

Committee on Equal Opportunities in Science and Engineering [CEOSE]. (2019). *Biennial Report to Congress 2017-2018: Investing in diverse community voices.* National Science Board, National Science Foundation. https://www.nsf.gov/od/oia/activities/ceose/reports/CEOSE_ReportToCongress_RP_FVmp_508.pdf

Dasgupta, N., & Stout, J. G. (2014). Girls and women in science, technology, engineering, and mathematics: STEMing the tide and broadening participation in STEM careers. *Behavioral and Brain Sciences, 1*(1), 21–29.

Davis, J., & Jett, C. C. (2019). *Critical race theory in mathematics education.* Routledge.

DeGruy, J. (2005). *Post traumatic slave syndrome: America's legacy of enduring injury & healing.* Upton Press.

Denner, J., Martinez, J., & Thiry, H. (2017). Strategies for engaging Hispanic/Latino youth in the US in computer science. In Y. Rankins & J. Thomas (Eds.), *Moving students of color from consumers to producers of technology* (pp. 24–48). IGI Global.

Frierson, H. T., Hood, S., & Hughes, G. B. (2002). Strategies that address culturally responsive evaluation. In. J. Frechtling (Ed.), *The 2002 user-friendly handbook for project evaluation* (pp. 63–73). National Science Foundation.

Gardner, T. Q., & Kowalski, S. E., & Kowalski, F. V. (2012, June), *Interactive Simulations Coupled with Real-time Formative Assessment to Enhance Student Learning* [Paper presentation]. ASEE Annual Conference & Exposition, San Antonio, TX, United States. doi: 10.18260/1-2—21583

Garrison, H. (2013). Underrepresentation by race-ethnicity across stages of U.S. science and engineering education. *CBE Life Sciences Education, 12*(3), 357–363.

Gay, G. (2010). *Culturally responsive teaching: Theory, practice and research* (2nd ed.). Teachers College Press.

Gonzalez, H. B. (2014). *The National Science Foundation: Background and selected policy issues.* (CRS 705700). Congressional Research Service.

Goodyear, L., Mansori, S., Ambat, E., & McMahon, T. (2019). *Building a STEM evaluation community: Evaluation and evaluation capacity building resource landscape study.* Education Development Center.

Graciá, H. A., Hotchkins, B. K., & McNaughtan, B. (2019). "Why not?" How STEM identity development promotes black transfer and transition. *The Journal of Negro Education, 88*(3), 343–357.

Gutiérrez, K. D., & Rogoff, B. (2003). Cultural ways of learning: Individual traits or repertories of practice. *Educational Researcher, 32*(5), 19–25.

Herman, M. (2004, May/June). Forced to choose: Some determinants of racial identification in multiracial adolescents. *Child Development, 75*(3), 730–748.

Hoegh, A., & Moskal, B. M. (2009, October). *Examining science and engineering students' attitudes toward computer science* [Paper presentation]. The 39th IEEE International Conference on Frontiers in Education, San Antonio, TX, United States. http://archive. fie-conference.org/fie2009/papers/1035.pdf

Hood, S., Frierson, H., & Hopson, R. (2005). *The role of culture and cultural context in evaluation: A mandate for inclusion, the discovery of truth and understanding.* Information Age Publishing.

Hood, S. H., Hopson, R. K., and Kirkhart, K. K. (2015). Culturally responsive evaluation: Theory, practice and future implications. In K. A. Newcomer, H. P. Hatry, & J. S. Wholey (Eds.), *Handbook of practical program evaluation,* (4th ed., pp. 281–317). Jossey-Bass.

Hopson, R. K. (2009) Reclaiming knowledge at the margins *culturally responsive evaluation* in the current *evaluation* moment. In K. Ryan & B. Cousins (Eds.), *International handbook of educational evaluation* (pp. 429–446). Sage.

Institute of Medicine. (2010). *Rising above the gathering storm, Revisited: Rapidly approaching category 5.* National Academies Press. https://doi.org/10.17226.12999

Ireland, D. T., Freeman, K. E., & Winston-Proctor, C. E. (2018). (Un)hidden figures: A synthesis of research examining the intersectional experiences of Black women and girls in STEM education. *Review of Research in Education, 42*(1), 226–154. https://doi. org/10.3102/0091732X18759072

Jacob, S., Nguyen, H., Togel-Grehl, C., Richardson, D., & Warschauer, M. (2018). Teaching computational thinking to English learners. *NYS TESOL Journal, 5*(2), 12–24.

Jordan, W. J. (2020). *Year 2 evaluation progress report: Strategies, the Bessie Coleman project, using computer modeling and flight stimulation to create STEM pathways.* Unpublished manuscript.

Joseph, N. M., & Cobb, F. (2019). Antiblackness is in the air: Problematizing Black students' mathematics education pathways from curriculum to standardized assessments. In J. Davis & C. C. Jett (Eds.), *Critical race theory in mathematics education* (pp. 140–163). Routledge.

Ketelhut, D. J. (2010). Assessing gaming, computer and scientific inquiry self-efficacy in a virtual environment. In L. A. Annetta & S. Bronack (Eds.), *Serious educational assessment: Practical methods and models for educational games, simulations and virtual worlds* (pp. 1–18). Sense Publishers.

Kim, A. Y., Sinatra, G. M., & Seyranian, V. (2018). Developing a STEM identity among young women: A social identity perspective. *Review of Educational Research, 88*(4) 589–625.

Koh, K. H., Basawapatna, A., Nickerson, H., & Repenning, A. (2014, July). Real time assessment of computational thinking. In *Proceedings of the IEEE Symposium on Visual Languages and Human-Centric Computing* (pp. 49–52). https://wiki.computationalthinkingfoundation.org/wiki/images/9/91/Paper_24.pdf

Ladson-Billings, G. (1995). Toward a theory of culturally relevant pedagogy. *American Educational Research Journal, 32*(3), 465–491.

Ladson-Billings, G. (2009a). *The dreamkeepers: Successful teachers of Black children.* Jossey-Bass.

Ladson-Billings, G. (2009b). Just what is critical race theory and what's it doing in a nice filed like education? In E. Taylor, D. Gillborn, & G. Ladson-Billings (Eds.), *Foundations of critical race theory in education* (pp. 17–36). Routledge.

Ladson-Billings, G. (2014). Culturally relevant pedagogy 2.0: a.k.a. the Remix. *Harvard Educational Review, 84*(1), 74–84.

Ladson-Billings, G., & Tate, W. F. (2006). Toward a critical race theory of education. In A. D. Dixson & C. K. Rousseau (Eds.), *Critical race theory in education: All god's children got a song* (pp. 11–30). Routledge.

Lakens, D. (2013). Calculating and report effect sizes to facilitate cumulative science: A practical primer for *t*-tests and ANOVAs. *Frontiers in Psychology, 4,* 863. doi:10.1016/j/cresp.2020.100002

Leinhardt, G., & Schwarz, B. B. (1997). Seeing the problem: An explanation from Polya. *Cognition and Instruction, 15*(3) 395–434.

Leonard, J. (2017). *Final project report to NSF: Visualization basics: Using gaming to improve computational thinking (uGame-iCompute).* Unpublished manuscript.

Leonard, J., Barnes-Johnson, J., Mitchell, M., Unertl, A., Stubbe, C. R., & Ingraham, L. (2017). Developing teachers' computational thinking beliefs and engineering practices through game design and robotics. In E. Galindo & J. Newton (Eds.), *Proceedings of the 39th annual meeting of the North American Chapter of the International Group for the Psychology of Mathematics Education* (pp. 1289–1296). Hoosier Association of Mathematics Teacher Educators.

Leonard, J., Buss, A., Gamboa, A., Mitchell, M., Fashola, O. S., Hubert, T., & Almughyirah, S. (2016). Using robotics and game design to enhance children's self-efficacy, STEM attitudes, and computational thinking skills. *Journal of Science Education and Technology, 28*(6), 860–876. doi:10.1007/s10956-016-9628-2

Leonard, J., Mitchell, M., Barnes-Johnson, J., Unertl, A., Outka-Hill, J., Robinson, R., & Hester-Croff, C. (2018). Preparing teachers to engage rural students in computational thinking through robotics, game design, and culturally responsive teaching. *Journal of Teacher Education, 69*(4), 386–407. doi:10.1177/0022487117732317

Leonard, J., Mitchell, M. B., Fashola, T. S., & Hubert, T. L. (2015). *Preparing teachers to engage in rural and indigenous students in computational thinking through game design* [Paper presentation]. Annual meeting of the American Educational Research Association, Chicago, IL, United States.

Margolis, J., Goode, J., & Flapan, J. (2017). A critical crossroads for Computer Science for All: "Identifying talent" or "building talent," and what difference does it make? In Y. Rankin & J. Thomas (Eds.), *Moving students of color from consumers to producers of technology* (pp. 1–23). IGI Global.

Margolis, J., Goode, J., & Ryoo, J. (2014). Democratizing Computer Science knowledge. *Educational Leadership, STEM for All, 72*(4), 48–53.

Martin, D. B., Gholson, M. L., & Leonard, J. (2010). Mathematics as gatekeeper: Power and privilege in the production of knowledge. *Journal of Urban Mathematics Education*, *3*(2), 12–24.

Masuoka, N. (2011). The "multiracial" option: Social group identity and changing patterns of racial categorization. *American Politics Research*, *39*(1), 176–204.

Mathison, S. (2008). What is the difference between evaluation and research – and why do we care? In N. L. Smith & P. R. Brandon (Eds.), *Fundamental issues in evaluation* (pp. 183–196). The Guilford Press.

McGee, E. O. (2018). "Black Genius, Asian fail": The detriment of stereotype lift and stereotype threat in high-achieving Asian and Black STEM students. *AERA Open*, *4*(4), 1–16.

McGee, E. O. (2020). Interrogating structural racism in STEM higher education. *Educational Researcher*, *49*(9), 633–644.

Mertens, D., & Hopson, R. (2006). Advancing evaluation of STEM efforts through attention to diversity and culture. *New Directions for Evaluation*, *109*, 35–51.

Mitchell, M., Fashola, O. S., & Leonard, L. (2016, April). *Broadening participation in rural settings* [Paper presentation]. Center for Culturally Responsive Evaluation and Assessment, Chicago, IL, United States.

Nasir, N. S., Atukpawu, G., O'Connor, K., Davis, M., Wischnia, S., & Tsang, J. (2009). Wrestling with the legacy of stereotypes: Being Black in math class. In D. B. Martin (Ed.), *Mathematics teaching, learning, and liberation in the lives of black children* (pp. 231–248). Routledge.

National Research Council (2010). *Report of a workshop on the scope and nature of computational thinking*. National Academy Press. https://doi.org/10.17226/12840

National Science Board (2018). *Science and Engineering Indicators 2018*. NSB-2018-1. National Science Foundation. https://www.nsf.gov/statistics/indicators/

National Science Foundation (2019). *Women, minorities, and persons with disabilities in science and engineering: 2019*. Special Report NSF 19-304. National Center for Science and Engineering Statistics. https://ncses.nsf.gov/pubs/nsf19304/

Newton, K. J., Leonard, J., Buss, A., Wright, C., & Barnes-Johnson, J. (2020). Informal STEM: learning with robotics and game design in an urban context. *Journal of Research on Technology in Education*, *52*(2), 129–147.

Noam, G., Shah, A. M., & Larson, J. D. (2014). Dimensions of Success Observation Tool. http://www.pearweb.org

Oakes, J., Joseph, R., & Muir, K. (2003). Access and achievement in mathematics and science: Inequalities that endure and change. In J. A. Banks & C. A. Banks, (Eds.), *Handbook of research on multicultural education* (2nd ed., pp. 69–90). Jossey-Bass.

Oakes, J., Ormseth, T., Bell, R., & Camp, P. (1990). *Multiplying inequalities: The effects of race, social class, and tracking on opportunities to learn mathematics and science* (R-3928-NSF). RAND.

O'Hara, R. M. (2020). STEM(ing) the Tide: A critical race theory analysis in STEM education. *Journal of Constructivist Psychology*. doi:10.1080/10720537.2020.184825.

Paris, D. (2012). Culturally sustaining pedagogy: A needed change in stance, terminology, and practice. *Educational Researcher*, *41*(3), 93–97.

Papazian, A. E., Noam, G. G., Shah, A. M., & Rufo-McCormick, C. (2013). The quest for quality in afterschool science: The development and application of a new tool. *Afterschool Matters*, *18*, 17–24.

Pearson, K. (2009). From a usable past to a collaborative future: Black culture in the age of computational thinking. *Black History Bulletin*, *72*(1), 41–44.

Pólya, G. (1945). *How to solve it: A new aspect of mathematical method.* Princeton University Press.

Powell, A., Nielsen, N., Butler, M., Buxton, C., Johnson, O., Ketterlin-Geller, L., Stiles, J., & McCulloch, C. (2018). *The use of theory in research on broadening participation in PreK-12 STEM education: Information and guidance for prospective DRK-12 grantees.* Education Development Center. https://cadrek12.org/sites/default/files/CADRE_BroadeningParticipationTheories.pdf.

Rankin, Y. A., Thomas, J. O., & Joseph, N. M. (2020). Intersectionality in HCI: Lost in translation. *Interactions, 27*(5), 68–71.

Rotter, J. B. (1990). Internal versus external control of reinforcement: A case history of a variable. *American Psychologist, 45*(4), 489–493.

Schoenfeld, A. H. (2018). Pólya, problem solving, and education. *Mathematics Magazine, 60*(5), 283–291.

Scriven, M. (1991). *Evaluation thesaurus* (4th ed.). Sage.

Siwatu, K. O. (2007). Preservice teachers' culturally responsive teaching self-efficacy and outcome expectancy beliefs. *Teaching and Teacher Education, 23*(7), 1086–1101.

Solorzano, D., Ceja, M., & Yosso, T. (2000). Critical race theory, racial microaggressions, and campus racial climate: The experiences of Black college students. *Journal of Negro Education, 69*(1/2), 60–73.

Stigler, J. W., & Hiebert, J. (1998). Teaching is a cultural activity. *American Educator, 22*(4), 4–11.

Stinson, D. W. (2004). Mathematics as 'Gate-keeper (?)': Three theoretical perspectives that aim toward empowering all children with a key to the gate. *Mathematics Educator, 14*(1), 8–18.

Tabb, K. M. (2015). Changes in racial categorization over time and health status: an examination of multiracial young adults in the USA. *Ethnicity & Health, 21*(2), 146–117.

Tabb, K. M., Larrson, C. R., Choi, S., & Huang, H. (2016). Disparities in health services using multiracial American young adults. *Journal of Immigrant and Minority Health, 18*(6), 1462–1469.

Thomas, V. G., & Campbell, P. B. (2021). *Evaluation in today's world: Respecting diversity, improving quality, and promoting usability.* Sage.

Unfried, A., Faber, M., Stanhope, D., & Wiebe, E. (2015). The development and validation of a measure of student attitudes toward science, technology, engineering and math (S-STEM). *Journal of Psychoeducational Assessment, 33*(7), 622–639.

U.S. Census Bureau (2017). *American Community Survey (ACS) demographic and housing estimates.* https://data.census.gov/cedsci/table?d=ACS%205Year%20Estimates%20Data%20Profiles&tid=ACSDP5Y2017.DP05.

Weiner, B. (1972). Attribution theory, achievement motivation, and the educational process. *Review of Educational Research, 42*(2), 203–215.

Weiner, B. (2010). Attribution theory. *International Encyclopedia of Education, 6*, 558–563.

Wilkins-Yel, K. G., Hyman, J., & Zounlome, N. O. O. (2019, August). Linking intersectional invisibility and hypervisibility to experiences of microaggressions among graduate women of color in STEM. *Journal of Vocational Behavior, 113*, 51–61.

Wing, J. M. (2006, March). Computational thinking. *Communications of the ACM, 49*(3), 33–35. https://doi.org//10.1145/1118178.1118215

Yadav, A., Mayfield, C., Zhou, N., Hambrusch, S., & Korb, J. T. (2014, March). Computational thinking in elementary and secondary teacher education. *ACM Transactions on Computing Education, 14*(1), 1–16. http://dx.doi.org/10.1145/2576872

APPENDIX A
COMPUTATIONAL THINKING RUBRIC

CT components	Emerging (1)	Moderate (2)	Substantive (3)
Formulating Problems	If–then statements unclear in terms of problem goals (e.g., "Can pigs fly?")	If–then statements create conditions that allow agent to move through program using a single condition (e.g., if you see a ghost move left)	If–then statements more complex and agent moves to more than one set of criteria (e.g., if you see a ghost and a scarecrow move to the left and/or up)
Abstraction	Agent and background resemble tutorial in students' game	Agent or background is non-traditional and created by the student	Agent and background are non-traditional and created by the student
Logical Thinking	If–then statements do not follow logical path (e.g., agent is stuck and cannot move through the program)	If–then statements follow logical path with some complexity (e.g., agent moves through the program but no real challenges)	If–then statements follow logical path with more complexity (e.g., agent moves through program but can run into danger)
Using Algorithms	No evidence of algorithmic use (i.e., game **cannot** keep score)	Some evidence of algorithm use (i.e., the game **can** keep score)	Evidence of algorithm use and final score: (i.e., the game keeps score and says "you won")

CT components	Emerging (1)	Moderate (2)	Substantive (3)
Analyzing & Implementing Solutions	No evidence of the ability to debug the program	Some evidence of debugging	Strong evidence of debugging
Generalizing and Problem Transfer	Game resembles tutorial	Game has some elements of tutorial but some differences	Game is not similar to tutorial at all and shows creative use of knowledge transfer
Use of Culture or Off-the-Shelf Games	No evidence of culture or inclusion of elements in off-the-shelf games	Some evidence of culture or similarities to off-the-shelf games with some modifications or improvements	Substantial cultural modeling or similarities to off-the-shelf games with improvements and/or significant modifications

APPENDIX B

ABBREVIATED BCP COMPUTER PROGRAMMING SELF-EFFICACY, COMPUTATIONAL THINKING, AND VALUE EXPECTANCY SURVEY

Construct	Cronbach's Alpha Reliability	Item
Computer Programming Self-Efficacy (5-point Likert scale: 1= strongly disagree to 5 = strongly agree)	α = 0.730	I am good at building computer programs. I am good at fixing computer programs. I am very good at building things in computer games. I can use a computer to control toys and tools.
Computational Thinking (5-point Likert scale: 1= strongly disagree to 5 = strongly agree)	α = 0.733	When solving problems, I can create a list of steps to solve it. When solving a problem, I can see patterns in the problem. When solving a problem, I can figure out several ways to solve it. When solving a problem, I can figure out what is the best solution.
Science Expectancy Value (7-point Likert scale: 1 = not good to 7 = very good or one of the worst to one of the best)	α = 0.840	How good at science are you? If you were to order all of the students in your class from the worst to the best in science, where would you put yourself? Compared to other students, how well do you expect to do in science this year? How well do you think you will do in your science course this year?

Construct	Cronbach's Alpha Reliability	Item
Technology Expectancy Value (4-point Likert scale: 1 = strongly disagree to 4 = strongly agree)	$\alpha = 0.822$	Working in technology is very creative. Technology is too difficult for me.* For students my age, technology is not interesting.* I am not interested in technology.* I feel comfortable working with computers. I'm relaxed when I work with a computer. I can do a good job using technology. I'm really good at using technology.

* Items were reverse coded.

APPENDIX C

BCP STUDENT INTERVIEW PROTOCOL

(1) Please tell us about yourself.
   (a) Your age, grade level
   (b) Reasons for joining the STEM club
(2) Tell us about the activity you did today.
   (a) What did you like?
   (b) What didn't you like about the activity?
   (c) How can we make the activity better?
(3) What did you learn about?
   (a) Drones?
   (b) Flight simulation?
   (c) Computer modeling?
(4) Tell us about your experiences with the field trip.
   (a) What did you like the most?
   (b) What suggestions do you have for us?
(5) Tell us about the guest speakers.
   (a) What did you learn?
   (b) What suggestions do you have for us?
(6) Anything else you'd like to share with us?

APPENDIX D

BCP MODULE: ARTIFACTS IN SPACE (GRADES 3–5)

Table of Contents

Lesson 1: Learning about Astronauts of Color and the Space Shuttle (One Day)

Essential Question: How can I increase students' knowledge about astronauts of color and expose them to careers in aerospace and engineering?

EXPLORE

Learning Objective:
• Students will be able to develop listening, reading, and comprehension skills by hearing and reading stories about contemporary astronauts and make a model of a space shuttle.

Vocabulary: Civil Rights, Black Lives Matter (#BLM), NASA, space shuttle, orbiter, astronaut

Activity 1: Who is Commander Victor Glover?

Read excerpts from the article on Commander Victor Glover and lead children into a discussion about the difficulty associated with becoming an astronaut. (See https://www.nasa.gov/astronauts/biographies/victor-j-glover/biography.)
After reading the article, answer the following questions:

- What assumptions do the authors of the text hold?
- What do you agree with in the text?
- What do you want to disagree with in the text?
- What parts of the text do you want to aspire to do?

Activity 2: Read Alouds and Independent Reading

- *Mae Among the Stars.* https://youtu.be/b_mfdqyBqT8 **and/or**
- *Ron's Big Mission.* https://youtu.be/RdPIx8JKuCs

Lead children into a brief discussion about one or both of the books and some of the assumptions made about Mae and/or Ron. Ask children what they aspire to be when they grow up. Talk briefly about what content knowledge they need to learn to follow their dreams.

Possible discussion questions:

1. What aspirations did Mae (Ron) have as a young child?
2. What preparations did Mae (Ron) make to accomplish her (his) dream?
3. What obstacles did Mae (Ron) have to overcome to make her (his) dream come true?
4. Who were the adults that helped Mae (Ron) realize her (his) dream?

Activity 3: Veggie Shuttle

Introduce the lesson by reading excerpts from the Space Shuttle Fact Sheet. (See https://www.nasa.gov/centers/johnson/pdf/167751main_FS_SpaceShuttle508c.pdf.) Then guide students through making a Veggie Shuttle from food items.
Materials: Carrots, celery, slice of bread, peanut butter or cheese spread (if allergic to peanut butter), paper plate, plastic knife. Other substitutes may include bread sticks instead of carrots.

MAKE IT WORK

Preparing for the Activity: Prepare carrots by having an adult slice them in half lengthwise. Cut lengthwise to make two large pieces with the thick end of the carrot at the top.

Directions for Veggie Shuttle:

Place one-half of the carrot piece on a paper plate, flat side down, to represent the external fuel tank (ET) of the Space Shuttle. This is the longest part of the Shuttle assembly. With peanut butter or cheese spread, attach two same-size celery sticks, one on each side of the carrot. These represent the solid rocket boosters (SRBs) of the Shuttle. They should be longer than the orbiter but shorter than the carrot. Place a paper template on top of the slice of bread and cut around it to make an orbiter. The orbiter should be shorter than the celery sticks. Spread peanut butter or cheese spread on the orbiter and attached it the carrot (ET). (See *Mission Mathematics II* [Hines & Hicks, 2005] for diagram and further instructions.)

Take a picture of your orbiter and simulate a launch sequence (10, 9, 8, 7. 6, … . 1—liftoff).

Simulate the launch and separate the SRB by detaching the two celery sticks and placing them on the paper plate. Students may eat the shuttle after taking a photograph of it.

Activity 4: A Day in Space

Launching the lesson: Students should brainstorm by making a list of things they think an astronaut might do in space. Encourage students to make a schedule that includes time for sleeping, meals, hygiene (brushing teeth, washing up, etc.), working, exercising, and relaxing each day. Astronaut work includes conducting experiments, putting on spacesuits to do extravehicular activities (EVA), building the ISS, cleaning the living area, and maintaining the orbiter. Next have the students discuss living quarters for four to seven astronauts to share. (See *Mission Mathematics II* [Hines & Hicks, 2005] for mid-deck floor plan.) Have students use a calculator to find the perimeter, area, and volume of the living space. Have them compare the space of the Shuttle's mid-deck to the space in their bedroom. How does volume make a difference in space?

SHARE

Students share their work with the teacher by uploading photos of their Veggie Shuttle.

Lesson 2: Culturally Relevant Artifacts for Space Launch (Two Days)

Essential Question: How can I increase students' spatial reasoning and computational thinking to scale a basket, gourd, or other artifact to launch into space?

EXPLORE

Learning Objective:
• Students will be able to develop listening, reading, and comprehension skills by learning how an Indigenous woman made a basket small enough to travel into space.

Vocabulary: loonie, planets, space station, scaling, weave, Tinkercad

Activity 1: Basket Goes into Space

Read excerpts from the article on Shanna Francis and lead children into a discussion about the difficulty associated with making very small objects. (See https://www.cbc.ca/news/canada/nova-scotia/canadian-astronaut-takes-mi-kmaq-basket-to-space-station-1.4929966.)

After reading the article, answer the following questions:

• What assumptions do the authors of the text hold?
• What do you agree with in the text?
• What do you want to disagree with in the text?
• What parts of the text do you want to aspire to do?

Activity 2: Read Alouds and Independent Reading

• *The Spider Weaver: A Legend of Kente Cloth.* https://www.youtube.com/watch?v=L-R9Wc855Xg

Lead children into a brief discussion about the book and some of the assumptions made about Kente cloth. Ask children what they may want to make as an artifact (basket or gourd) that can be 3D printed to carry into space.

Possible Discussion Questions:

1. In what country did the story about the spider weaver take place?
2. What does a weaver do?
3. How were the weavers inspired to make Kente cloth?
4. What designs and patterns did you notice in the Kente cloth the weavers created?

MAKE IT WORK

Activity 3: Artifact in Space

Students will use the Tinkercad tutorial (see https://osf.io/fgyh6/) to create a basket or gourd that Commander Victor Glover could take into space on his next mission.

Preparing for the Activity: Students will need an email account to log onto Tinkercad. See https://www.tinkercad.com to guide students in making the basket or gourd.

SHARE

Students should save Tinkercad files and share them with the teacher and ITEST staff for 3D printing. A contest will take place to judge the best artifact.

Students may also choose to write letters to Commander Glover to explain why he should select their artifact for a future space mission.

Lesson 3: Mission Mars (Two Days)

Essential Question: How can I increase students' knowledge of the Solar System, including the planets?

EXPLORE

Learning Objective:

Students will be able to develop listening, reading, and comprehension skills about the Mars Reconnaissance Orbiter (MRO). Participants will also learn about the characteristics of the eight planets in the Solar system and how far each planet is from the Earth.

Vocabulary: planet, dwarf planet, probability, experiment, orbiter

Activity 1: Introduce vocabulary and have children watch Youtube video clip of women working for NASA in the movie *Hidden Figures*: https://youtu.be/PcKzIDc1V-s

- What assumptions do the authors of the clip hold?
- What do you agree with in the clip?
- What do you want to disagree with in the clip?
- What parts of the clip do you want to aspire to do?

Activity 2: Mission to Mars
Goals of the lesson:

- To play a game that models a uniform probability distribution
- To record and organize cumulative data from repeated probability experiments
- To display and analyze data to identify a pattern
- To identify events that are equally likely

Materials: Brief article about Mars Reconnaissance Orbiter (MRO): https://www.jpl.nasa.gov/missions/mars-reconnaissance-orbiter-mro/. Resource page on Mission to Mars (six copies), computer or phone for random number generator, and crayons. (See *Mission Mathematics* II [Hines & Hicks, 2005].)

Launching the Activity: After sharing the background information about MRO, explain to students that they are going on a journey to that planet. Because the planet is at a long distance, students need to find which of the six spacecraft will travel there the fastest. Students can play as individuals or with a partner. Ask the child to predict which spacecraft will be the winner at the beginning of the game.

Directions:

- Set the random number generator to select no more than six digits. Press the generator to get a number. Alternatively students could roll a die. Make note of the number that comes up on the generator or the die.
- On the first gameboard, color a square above the spaceship that represents that number.
- Continue in this manner until one column is completely colored. This spacecraft is the winner!
- Play the game several times to see if you get the same result.

Discussion: Ask the child(ren) to explain how different spacecraft can win if the game is played repeatedly. They should discover that each spacecraft is equally likely to win due to the fact that each number has an equal chance of coming up on the random number generator or die.

Activity 3: Are We There Yet? How Long Will It Take?

Students will use calculators (cell phone or computer calculator may be used) and the planet worksheet to determine how long it would take to travel to each of the planets, including the dwarf planet, Pluto. (See *Mission Mathematics II* [Hines & Hicks, 2005].)

For example, the distance from Earth to Venus is 26,000,000 miles. The speed of the average shuttle was 17,500 miles per hour. The approximate number of hours (round to nearest whole number) it would take to fly from Earth to Venus is (26,000,000 m /17,500 m/h) 1485 hours. There are 24 hours in a day. To find the number of days (round to nearest whole day): 1485 h /24 h/d = 62 days. This is less than one year (about 2 months) since there are 365 days in a year.

After calculating the distances and time (hours, days, years) of travel, students will complete a worksheet (see *Mission Mathematics II* [Hines & Hicks, 2005]) to explain which planet they would like to visit and why.

SHARE

Students should save and submit worksheets to the teacher.

ENRICHMENT

If time permits and students have an interest, they can develop a SCRATCH game (see tutorial on pp. 182-187) about space. Themes may include creating spaceships, astronauts, aliens, etc.

https://scratch.mit.edu

RESOURCES

Ahmed, R. (2018). *Mae Among the Stars*. HarperCollins.

Blue, R., & Naden, C. J. (2009). *Ron's Big Mission*. Dutton Books (Penguin Group).

Hines, D., & Hicks, D. (2005). *Mission Mathematics II: Grades 3–5*. NCTM.

Leonard, J., & Oakley, J. E. (2018). We have lift off! Integrating space science and mathematics into elementary classrooms. *Journal of Geoscience Education, 54*(4), 452–457.

Musgrove, M. (2001). *The Spider Weaver: A Legend of Kente Cloth*. The Blue Sky Press.

STANDARDS

NCTM STANDARDS http://www.nctm.org	NGSS STANDARDS	SOCIAL STUDIES STANDARDS	WIDA STANDARDS https://wida.wisc.edu/teach/standards/eld
Number and Operations—understand place value and base ten number system, operations (+, -, ×, /) whole numbers and decimals **Probability**—understand chance and likelihood of events **Algebra**—describe, extend, and make generalizations about geometric and numeric patterns; represent and analyze patterns and functions, using words, tables, and graphs; model problem situations with objects and use representations, such as graphs, tables, and equations to draw conclusions **Problem solving**—find solutions to problems; debugging Tinkercad or SCRATCH program **Measurement**—understand attributes such as length and speed; develop strategies for estimating perimeter, area, and volume of irregular shapes; select and use benchmarks to estimate measurement **Geometry**—learning to use spatial figures to create objects; draw and build geometric objects; identify and build 3D objects in 2D space	**Earth and the Solar System** 5-PS2-1. Support an argument that the gravitational force exerted by Earth on objects is directed down [and lack of gravity in space allows for more use of space aboard ISS]. 5-ESS1-1. Support an argument that differences in the apparent brightness of the sun compared to other stars [and planets] are due to their relative distances from Earth. 5-ESS1-2. The orbits of Earth around the sun and of the moon around Earth, together with the rotation of Earth about an axis between its North and South poles, cause observable patterns [leading to calculations of days, months and years].	**Culture—Human beings create, learn, share, and adapt to culture.** Students examine the socially transmitted beliefs, values, institutions, behaviors, traditions, and way of life of a group of people; it also encompasses other cultural attributes and products, such as language, literature, music, arts and artifacts, and foods. **Science, Technology, & Society**—Children learn how science and technologies influence beliefs, knowledge, and their daily lives.	**Social and instructional language**—reflects the ways in which students interact socially to build community and establish working relationships with peers and teachers in ways that support learning. **Language of Language Arts**—English language learners communicate information, ideas and concepts necessary for academic success in the content area of language arts. **Language of Mathematics**—English language learners communicate information, ideas, and concepts necessary for academic success in the content area of mathematics.

SCRATCH: DANCING ON THE MOON

TUTORIAL

The SCRATCH interface is split into four main sections: Main Window, Sprite Window, Scripts Area, and Block Area. The Main Window is where the results of your block sequence are shown. The Sprite window displays the selectable sprites in your project. The Block Window is where you select various blocks for the project, and the Script Area is where you build the project.

STEP 1. Go to SCRATCH website and click **start creating**. Minimize the pop-up tutorial.

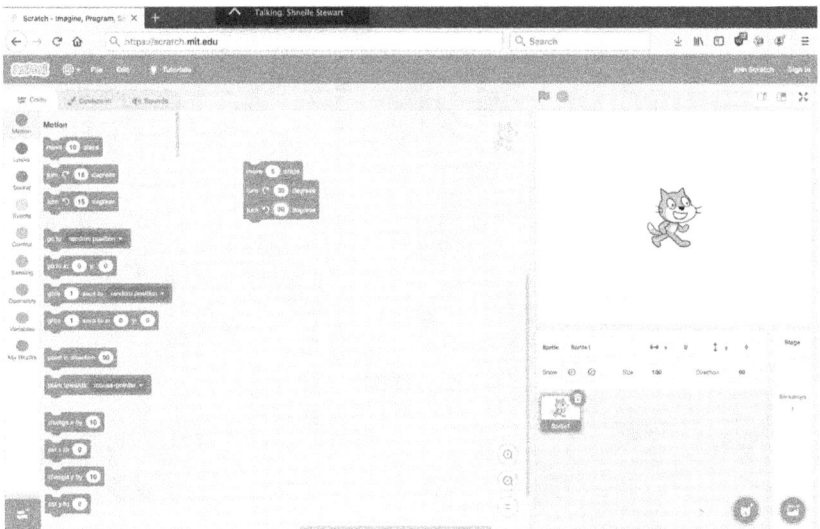

STEP 2. The figure in STEP 1 shows three move blocks were clicked-and-dragged to make the Sprite move across the screen in the Main Window. Double-clicking the number in the middle of the block will allow you to change the number of steps the cat will take.

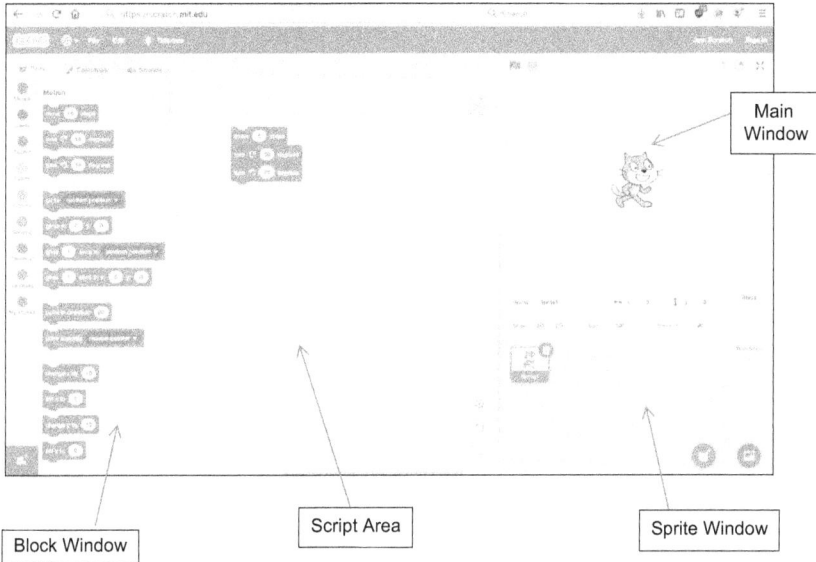

Main Window

Block Window

Script Area

Sprite Window

STEP 3. In the Block Window, click on "Sound" to bring up the sound block options. Click and drag a PLAY SOUND block and snap it under the MOVE block. Additional move and sound blocks may be added. You must also select a block to turn off the sound.

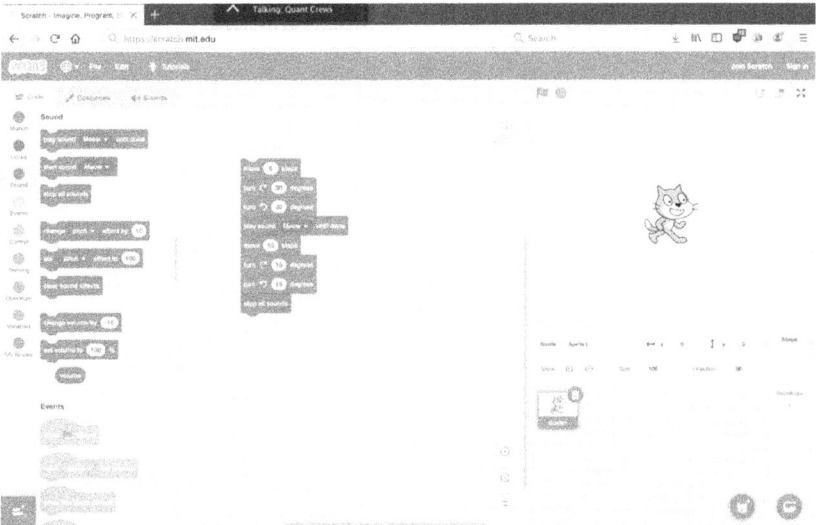

STEP 4. In the Block Window, click "Control" and drag out a REPEAT block. Snap it around the entire block stack.

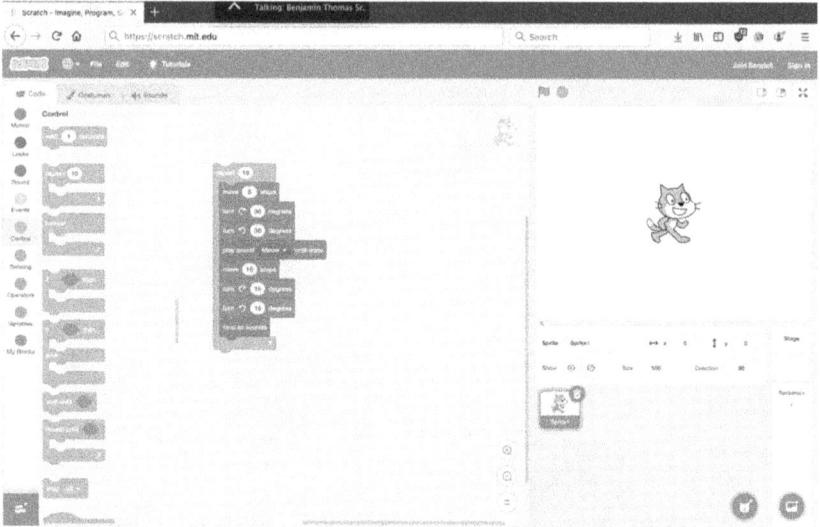

STEP 5. In the Block Window click "Looks" and drag out the "SAY _____ FOR __SECS" block. In the text box, type whatever words you desire your cat to say. Snap this block to the top of the stack. Your cat will say whatever phrase you typed in the text box before it begins to move.

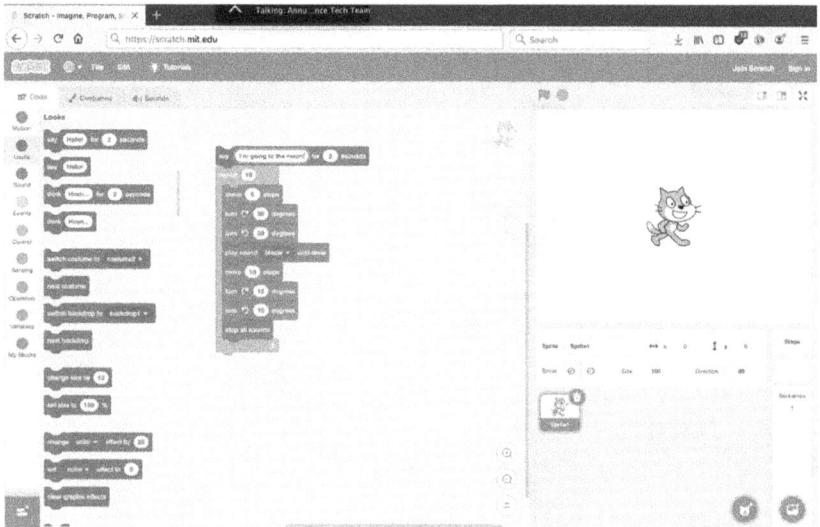

STEP 6. In the Block Window, click "Events." Drag out a WHEN (GREEN FLAG) CLICKED block and snap it to the top of the stack. This will allow you to start your event by simply clicking the green flag in the top right corner of the Main Window.

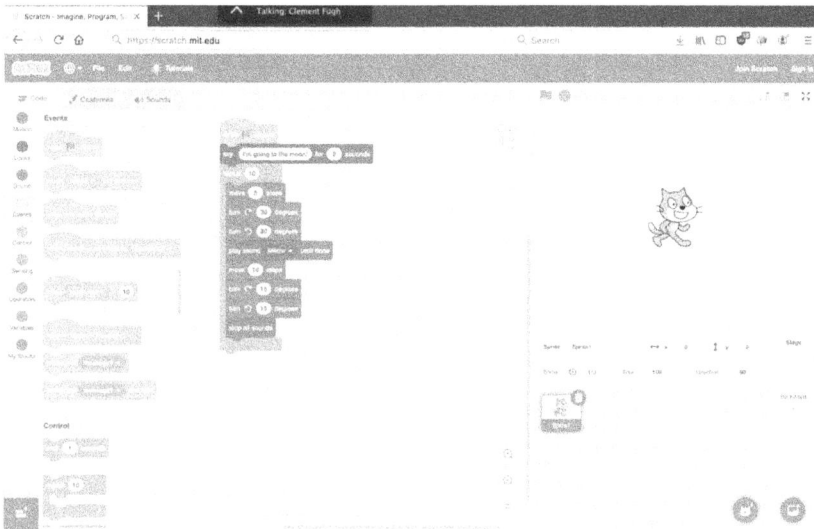

STEP 7. Now we have to add a backdrop to our game. On the left side of the Sprites Window there is button with a photo icon. This button allows you to select a backdrop from the library or upload one from the computer. Click this button and find the image you want for your game.

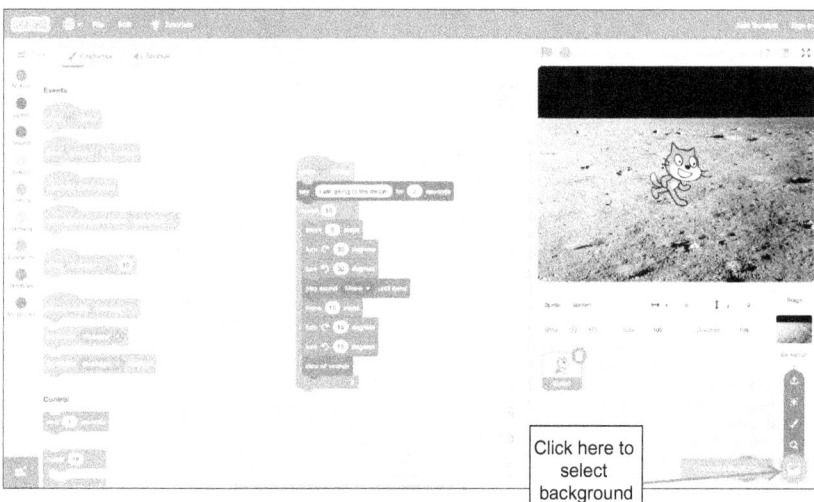

Click here to select background

STEP 8. Once you have the background you can add additional sprites by clicking the sprite icon. Ripley was added from the SCRATCH library.

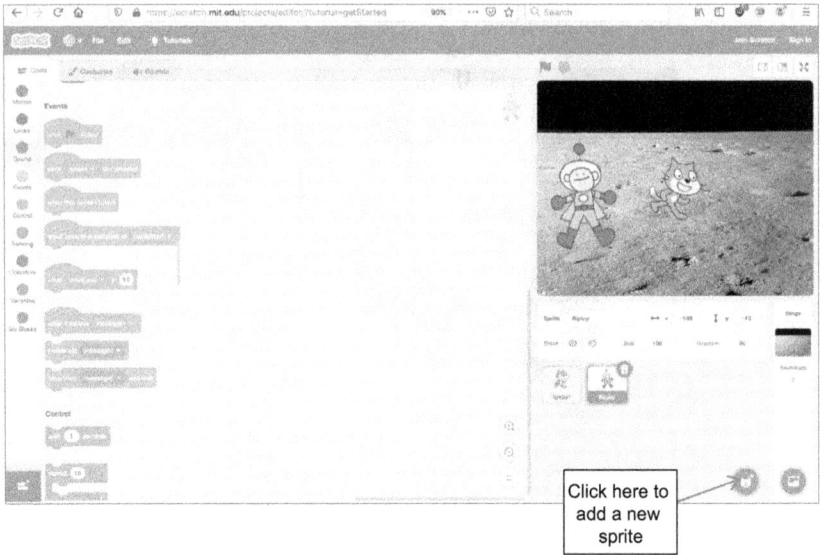

STEP 9. The final step is to press the green flag for the Sprites to dance on the moon.

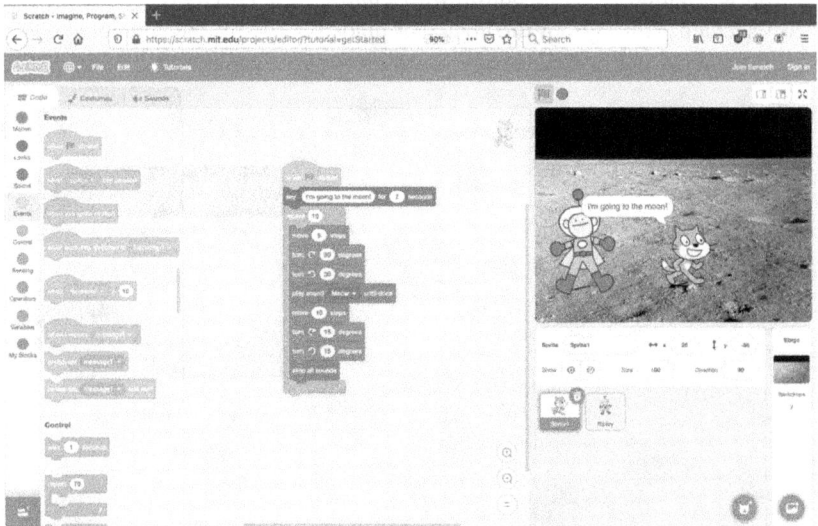

STEP 10. The program runs for both Ripley and the cat.

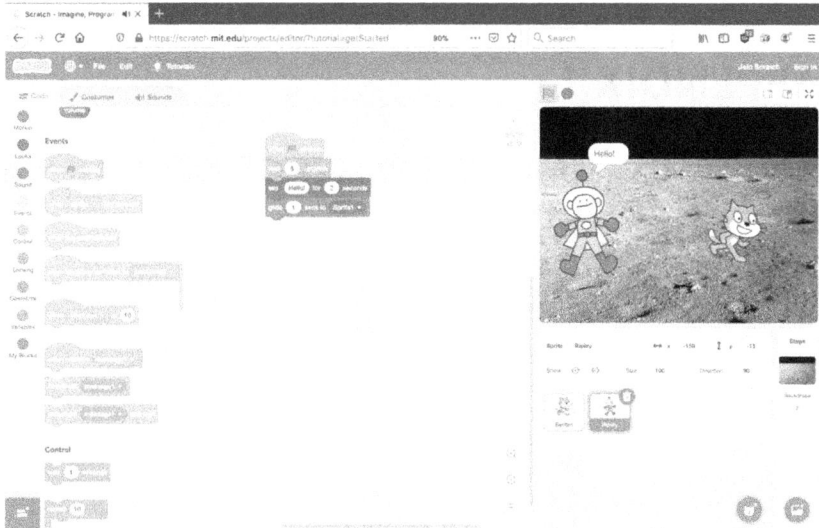

APPENDIX E

IEK STEM SUMMER CAMP LIST OF ACTIVITIES, JUNE 24–28, 2019

Time	Monday	Tuesday	Wednesday	Thursday	Friday
9:30am–10:30am	Group A/B Intro/ Pre-Surveys	Group A/B Iva's Sister	Field Day	Group A/B Activity/Guest	Group A/B Activity/ Guest
10:30am–12:00pm	Group A Topo Maps	Group A CNC Maps		Group A 3D Printing	Group A/B
					Logistics and Post-Surveys
	Group B 4 Forces/Drone Safety/Fly	Group B GIS/Flight Modes/Fly	Data Collection Area Transects	Group B Digital Maps	
			Aerial Footage of Riparian		
12:00pm–12:30pm	Lunch	Lunch		Lunch	Lunch
12:30pm–2:00pm	Group A 4 Forces/Drone Safety/Fly	Group A GIS/Flight Modes/Fly	Buffalo Jumps/Tipi Rings/etc.	Group A Digital Maps	Group A/B Student Presentations

Time	Monday	Tuesday	Wednesday	Thursday	Friday
	Group B Topo Maps	Group B CNC Maps		Group B 3D Printing	History Moment Bessie Coleman
2:00pm– 2:30pm	Group A/B Wrap Up and Dismissal	Group A/B Wrap Up and Dismissal		Group A/B Wrap Up and Dismissal	Group A/B Wrap Up and Dismissal

APPENDIX F

BCP TEACHER INTERVIEW PROTOCOL

Topic domains for each question are indicated in bold. Non-bold language below each topic domain represents the lead question and follow-up prompts to enable additional elaboration and depth of response.

1. **Understanding Computational Thinking (CT)**
    How would you explain what computational thinking (CT) is? Prior knowledge of CT before PD? Probe reasons for limited prior exposure. Relevance for teaching mathematics and science today.

2. **Professional Development provided by project team**
    What types of professional development (PD) have you participated in as part of the project? How is it different than PD you've participated in before? What aspects of this PD have been most helpful to you? In what ways? What are the most important things you've learned from this PD? Why is that important to you?

3. **Increased Teacher Knowledge—Computational Thinking (CT)**
    Has the PD been effective in helping you learn more about CT? If so, how? If not, is there anything about the PD that interferes with or limits your opportunity to learn more about CT?

    (If CT learning occurred:) What have you learned about CT that has been helpful to you? What aspects of the PD most contributed to your growth in knowledge about CT?

    (If CT learning did not occur:) What can be done better in the PD to help you learn more about CT?

4. **Increased Teacher Knowledge-Culturally Responsive Pedagogy (CRP)**

   Has the PD been effective in helping you learn more about CRP? If so, how? If not, is there anything about the PD that interferes or limits your opportunity to learn more about CRP?

   (If CRP learning occurred:) What have you learned about CRP that has been helpful to you? What aspects of the PD most contributed to your growth in knowledge about CRP?

   (If CRP learning did not occur:) What can be done better in the PD to help you learn more about CRP?

5. **Application to Classroom Practice**

   How has the PD been relevant to your classroom instruction? What aspects of the PD have you incorporated into instruction? How do you now address CT in your instruction? How do you integrate CRP into your teaching?

6. **Change in Classroom Practice**

   When you think about changes you've made in your teaching as a result of the PD, what has been the most effective? What has been difficult? What other information or support do you need to continue the process of addressing CT and implementing CRP into your teaching?

   Are there any classroom realities that should be addressed or considered by the PD to help you improve in integrating CT and CRP into your teaching?

7. **Student Learning**

   How has student learning been affected by some of the changes you've introduced into your teaching as a result of the PD? How has your approach to CT affected student learning? Effect of CRP on student learning?

8. **Additional Information**

   Provide opportunity for participants to make other comments or information related to discussion.

INDEX

Page numbers in **bold** indicate tables, page numbers in *italics* indicate figures and page numbers followed by n indicate notes.

For Product Safety Concerns and Information please contact our EU
representative GPSR@taylorandfrancis.com
Taylor & Francis Verlag GmbH, Kaufingerstraße 24, 80331 München, Germany